The Numerical Solution of Algebraic Equations

R. Wait

Department of Computational and Statistical Science
University of Liverpool

A Wiley–Interscience Publication

JOHN WILEY & SONS
Chichester · New York · Brisbane · Toronto

Library of Congress Cataloging in Publication Data:

Wait, R.
The numerical solution of algebraic equations.

'A Wiley–Interscience publication.'
Includes index.
1. Equations–Numerical solutions. I. Title.
QA218.W34 519.4 78-21869
ISBN 0 471 99755 2

Typeset by Preface Ltd., Salisbury, Wilts.
and printed in Great Britain by
Unwin Brothers Ltd., Woking, Surrey.

To
Graeme and Jennifer

Preface

This book is intended as a half-year course for undergraduate students of mathematical science and is based on a one semester course given to second-year students at the University of Liverpool. The object of such a course is to provide guidance to those using or writing computer programs or intending to do so. At the same time it is an attempt to provide an insight into the mathematical analysis for those students with no more than a passing interest in the computational aspects of the methods. The individual sections are roughly equal in length and content, and they could each be reasonably covered in a single lecture. The sections can, in most cases, be classified as either theoretical or practical and there are approximately equal numbers of each.

The book is in two parts and, with one or two exceptions, there is no specific cross referencing, so the two parts can be read in either order. There are a few sections towards the end of Part II which, in view of their somewhat limited appeal, can be safely omitted.

In Part II special methods for polynomial equations are omitted as they have been dealt with effectively elsewhere and it would be difficult to improve on the presentation in, for example, Householder (1970). In addition, the usefulness of such methods in practice is highly questionable, since a polynomial equation can be formulated as a matrix eigenproblem and then solved by the QR algorithm. In the solution of arbitrary non-linear problems, there is no attempt to emulate the breadth and rigour of advanced texts such as Traub (1964), Ostrowski (1966), Ortega and Rheinboldt (1970) or Rheinboldt (1974). There is, however, an attempt to provide a thorough introduction to the theory of iterative methods such as cannot be found in general introductory texts on numerical analysis, which usually contain a very sketchy treatment of this area. At the same time numerous exercises and worked examples illustrate the directions in which the analysis proceeds in more advanced situations. It is intended herein to go beyond the ubiquitous Newton's method and give practical methods that work in practical problems, together with some indication of the relevance of current theory to current practice.

A knowledge of the calculus as far as Taylor series expansions is assumed, as is a degree of familiarity with the concepts of vectors and matrix norms. In order to avoid the book becoming a mere catalogue of all the available methods it has been necessary in certain instances to omit any discussion of some of the alternatives in order to concentrate on one of the best. Thus there is no explanation of Householder transformations because there is a careful assessment of Givens transformations: in particular there is one of the first accounts at this level, of the square-root-free Givens transformations. No attempt has been made to discuss the associated subjects of matrix eigenproblems, linear programming and optimization, although their connexions with the solution of algebraic equations is indicated where appropriate.

This is one of the first introductory text books (as opposed to a specialist monograph or the proceedings of a conference) to include a discussion of many of the practical difficulties to be encountered when solving systems of equations and to describe some of the possible methods of overcoming them. In particular, there is a discussion of sparse-matrix techniques for linear systems (the cases of an irregular pattern and of a regular structure are both covered), Brown's Newton–Gauss–Seidel type of method for systems, quasi-Newton methods for non-linear systems and orthogonal factorization of matrices, together with a description of their practical application in the numerical solution of equations. Robust methods for single equations are covered; in particular the Pegasus method appears in a textbook for the first time. All these methods are used by practical people to solve practical problems and students should therefore be aware of them at an early stage. Such methods should be included in any book that claims to show how to solve problems; up to the present these methods have in general been restricted to research papers in the scientific journals and an elementary exposition is long overdue.

It is to be expected that practising scientists and engineers will find this book of value as an introduction to the modern methods that are available. It would be possible for them to proceed to a more detailed account by following up the references in the extensive bibliography provided. It should also prove to be a suitable textbook for many of the short courses in numerical methods that now form part of a typical course for engineering students. On the other hand, students with a more mathematical bias should find sufficient material to stimulate their particular interests in what is, without doubt, one of the key foundations of numerical mathematics, on which other areas – numerical solution of differential equations for example – can build.

I would like to acknowledge the assistance that I received from colleagues and students, past and present, in preparing the manuscript of this book. In particular J. M. Watt and N. G. Brown deserve a special mention. In addition I would like to thank the secretaries of the department of Computational and Statistical Science at the University of Liverpool for their painstaking typing of large parts of the manuscript.

Liverpool
July 1978

R. Wait

Contents

PART II NON-LINEAR EQUATIONS

*Can be omitted on first reading.

1

Introduction

1.1 Linear Equations: An Introduction

One of the most fundamental problems of numerical analysis is the solution of a system of linear algebraic equations. Another cornerstone of the subject – the solution of non-linear equations – is also dealt with.

The numerical solution of linear algebraic systems is the object of a considerable amount of ongoing study, primarily in the solution of large sparse systems (see for example Rose and Willoughby, 1972; Bunch and Rose, 1976) and in the production of computer library packages, of which the most widely used are the I.M.S.L. Library, the FUNPACK, EISPACK, and LINPACK routines of the Argonne Laboratory, and the N.A.G. Library. This book will however be concerned only with a few of what might reasonably be termed *basic* methods and as much (or as little) theory as is considered appropriate. There is no attempt to provide an exhaustive catalogue of methods – merely enough to suggest the breadth of the subject. Whilst it is intended, as far as possible, to provide efficient and up-to-date methods, it should be remarked that it is on the computer implementation that such judgements ought to be made, and so without lengthy computer programs any assessments may of necessity be extremely vague and have limited validity. The important distinction between the mathematical statement of a method and the computer implementation should never be overlooked, and it is no coincidence that it is discussed again.

Classification

There are three largely distinct groups of methods:

(1) direct methods;
(2) iterative methods;
(3) other types.

The third group consists of such methods as *conjugate gradients* – see for example Engeli, Ginsburg, Rutishauser and Stiefel (1959) for a rather elegant account, or Reid (1971b) for a more up-to-date approach. These methods have their uses in various applications but unfortunately cannot find a place in the present slim volume. The sections on direct methods are not only important as a summary of some of the basic methods for solving linear equations, but also provide an introduction to the concepts of *matrix factorization*, which has many other applications. One very important application is the solution of the *algebraic eigenvalue problem*, viz. for any matrix A determine non-trivial $\mathbf{x}$ and λ such that

$$A\mathbf{x} = \lambda\mathbf{x}.$$

Several excellent texts are available that consider this problem – for example Wilkinson (1965a) and Gourlay and Watson (1973). The section on iterative methods for linear systems is delayed until a more appropriate place in Section 5.5 alongside descriptions of iterative methods in general, but it can be read out of sequence after Chapter 2.

1.2 Iterative Methods: An Introduction

Any iterative method consists of three parts:

(1) an initial estimate (or set of estimates) of the solution;
(2) a formula for updating the approximate solution;
(3) a 'fail-safe' procedure for stopping the updating process.

The different components and their relative importance are emphasized when the iterative process is represented by a flow chart such as Figure 1.

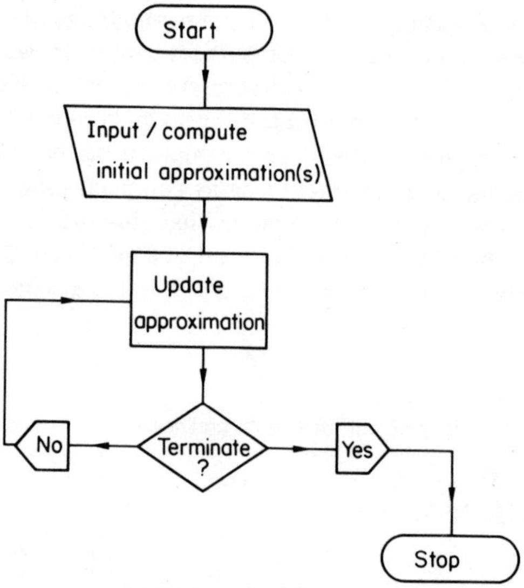

Figure 1 The iterative process

It is important to distinguish between the *iterative algorithm* – the total process om *start* to *stop* – and a single part of that process namely, the iteration formula. o be precise, mathematical (or non-mathematical) statements of (1), (2), (3) do ot constitute an algorithm; it is only a *computer implementation* of the method ıat can justifiably be termed an algorithm. There are usually many different ways f expressing a particular mathematical formula, but they are rarely equally suitable a basis for numerical computation.

A classical example of the difference between 'suitable' and 'unsuitable' ıathematical formulae arises in computing the roots of the quadratic equation

$$ax^2 - 2bx + c = 0.$$

he roots can be expressed as either

$$\frac{b \pm \sqrt{(b^2 - ac)}}{a} \qquad \text{or} \qquad \frac{c}{b \mp \sqrt{(b^2 - ac)}}.$$

/hen $|ac| \ll b^2$, b and $\sqrt{(b^2 - ac)}$ are approximately equal in magnitude. Since ıbtracting nearly equal numbers invariably results in a loss of accuracy in the ifference, the roots should be computed as

$$\frac{b + \sqrt{(b^2 - ac)}}{a}, \qquad \frac{c}{b + \sqrt{(b^2 - ac)}} \geqslant 0,$$

f $b < 0$, the minus sign is taken. This particular device for reducing *round-off rrors* appears again in Sections 5.7 and 6.2.

The terminating condition is always of vital importance and it must anticipate all ossible outcomes of the updating process. The method may not work every time nd so the algorithm should be able to judge when the method has failed. Since teration provides a sequence of approximate solutions, the algorithm will reach a uccessful conclusion when it adjudges the most recent approximation to be ufficiently accurate. An additional test in the algorithm should decide when the teration has spent enough time/money even though the desired accuracy has not een attained. It should be noted that the first stage often needs to include more han a straightforward input of data, it may incorporate a test to ensure that the nitial data are compatible with the update formula; for example, the iteration may equire two numbers x and y such that $x \leqslant y$, in which case it is essential to check he ordering of the input data.

In general in this book, it is the updating section of the algorithm that is liscussed. All the update formulae considered in Chapter 5 require one approximate olution, invariably denoted by $x^{(n)}$ ($\mathbf{x}^{(n)}$ for a system of equations), and generate further approximation $x^{(n+1)}$ ($\mathbf{x}^{(n+1)}$). It is to be hoped that $x^{(n+1)}$ is a better pproximation than $x^{(n)}$.

Examples of iteration formulae

(1) For finding $a^{1/2}$ for any positive number a; that is for solving the equation

$x^2 - a = 0$:

$$x^{(n+1)} = \frac{x^{(n)2} + a}{2x^{(n)}}$$

or

$$x^{(n+1)} = \frac{x^{(n)}(3a - x^{(n)2})}{2a}.$$

Note that both formulae have the solution $x^{(n)} = x^{(n+1)} = a^{1/2}$.

(2) For solving an arbitrary non-linear equation in the form $f(x) = 0$:

$$x^{(n+1)} = x^{(n)} - \left\{ \frac{\mathrm{d}f(x^{(n)})}{\mathrm{d}x} \right\}^{-1} f(x^{(n)}). \tag{1.1}$$

(3) For solving a system of linear equations $A\mathbf{x} = \mathbf{b}$, where $A = \{a_{ij}\}$,

$$\mathbf{x} = [x_1, \ldots, x_N]^{\mathrm{T}} \qquad \text{and} \qquad \mathbf{b} = [b_1, \ldots, b_N]^{\mathrm{T}},$$

that is the system of equations

$$\sum_{j=1}^{N} a_{ij}x_j = b_i, \qquad a_{ii} \neq 0, \qquad i = 1, \ldots, N:$$

$$\mathbf{x}^{(n+1)} = [x_1^{(n+1)}, \ldots, x_N^{(n+1)}]^{\mathrm{T}}$$

is given in terms of

$$\mathbf{x}^{(n)} = [x_1^{(n)}, \ldots, x_N^{(n)}]^{\mathrm{T}}$$

as

$$x_i^{(n+1)} = \frac{-1}{a_{ii}} \left(\sum_{\substack{j=1 \\ j \neq i}}^{N} a_{ij}x_j^{(n)} - b_i \right), \qquad i = 1, \ldots, N.$$

Exercise

1.1 Write out the formula (1.1) in the particular cases

(a) $f(x) = x^2 - a, \qquad a > 0$

(b) $f(x) = \frac{1}{x} - a.$

1.3 Algebraic Problems: An Introduction

The types of equations illustrated in the previous section arise naturally in the formulation of mathematical models. A few such models are given in this section to emphasize the importance of this area of numerical analysis.

(a) Chemical equilibrium problems

In simple models of chemical reactions, if it is assumed that all chemicals are kept well mixed, then spatial differences can be ignored. If, in addition, it is assumed that the equilibrium state has been reached, then transient effects can be ignored. The resulting conditions on the equilibrium state reduce to a set of non-linear algebraic equations.

As an example, consider the problem of finding the acidity of a solution made from a little soda lime and a lot of rhubarb leaves, that is a solution of oxalic acid $(COOH)_2$, and sodium hydroxide, NaOH. If we denote by [A] the molecular concentration of substance A, the equilibrium equations are conservation of mass and balance of electric charge in the solution.

The oxalic acid partially dissociates, so that

$$[(COOH)_2]_{total} = [(COOH)_2] + [H(COO)_2^-] + [(COO)_2^{2-}] \tag{1.2}$$

for which the dissociation constants are known to be

$$K_1 \equiv \frac{[H^+]\,[H(COO)_2^-]}{[(COOH)_2]} \tag{1.3}$$

and

$$K_2 \equiv \frac{[H^+]\,[(COO)_2^{2-}]}{[H(COO)_2^-]}\,. \tag{1.4}$$

If it is a weak solution, then the sodium hydroxide dissociates completely, so that

$$[NaOH]_{total} = [Na^+]\,.$$

Since the change is balanced, it follows that

$$[Na^+] + [H^+] = [OH^-] + [H(COO)_2^-] + 2[(COO)_2^{2-}]\,. \tag{1.5}$$

The dissociation constant for water is known to be

$$K_3 \equiv [H^+]\,[OH^-]\,.$$

Using (1.2)–(1.4) to eliminate the unknowns $[(COOH)_2]$, $[H(COO)_2^-]$ and $[(COO)_2^{2-}]$ from (1.5) leads to a quartic equation for $x \equiv [H^+]$ in terms of the known values

$$K_1, K_2, K_3, K_4 \equiv [(COOH)_2]_{total} \qquad \text{and} \qquad K_5 \equiv [NaOH]_{total}$$

in the form

$$f(x) \equiv (x^2 + xK_5 - K_3)(x^2 + xK_1 + K_1K_2) - x(K_1x + 2K_1K_2)K_4 = 0.$$

It is known that $[H^+] \geqslant 0$ and $[H^+] \leqslant (K_3)^{1/2}$, so once the values of $K_1, \ldots, K_5$ are determined it is possible to use the computer program in Section 6.4 to solve this problem.

(b) Leontieff input–output analysis

An economic model reduces to a system of algebraic equations only when the economy is assumed to be in equilibrium. The Leontieff analysis determines the level of production in an economy with many sections, each buying goods from the others in order to produce their own end products. The model could be of the national economy or of a single large industry with many divisions buying and selling amongst themselves as well as to the general public.

Let x_{ij} be the volume of sales from industry i to industry j. Let x_i be the gross output of industry i and let b_i be the external consumer demand for the products of industry i. Thus it follows that if there are N industries in the model, then

$$\sum_{j=1}^{N} x_{ij} + b_i = x_i \qquad i = 1, \ldots, N.$$

Assume that the sales from industry i to industry j are a fixed multiple of the production of industry j; then

$$x_{ij} = c_{ij}x_j, \quad i, j = 1, \ldots, N.$$

This situation would occur if industry i provided components for each product of industry j. The model then leads to

$$\sum_{j=1}^{N} a_{ij}x_j = b_i, \qquad i = 1, \ldots, N,$$

where all the

$$a_{ij} = \begin{cases} -c_{ij}, & i \neq j \\ 1 - c_{ii}, & i = j \end{cases}$$

and b_i ($i = 1, \ldots, N$) are known.

If it is a model of a particular sector of the economy with some industries producing components for other industries, the matrix $A = \{a_{ij}\}$ is upper triangular, that is

$$a_{ij} = 0, \qquad i > j.$$

Alternatively if the model is of a very large system such as the national economy, there will be a large number of zeros corresponding to sections that have no direct trade. In this case the matrix will be large and sparse and a suitable candidate for the methods of Section 4.2.

(c) General market equilibrium

Assume that the supply and demand for each of the products $y_1, \ldots, y_N$ are not fixed but are governed by the price of the product and of its competitors. Let S_i, D_i and P_i be respectively the supply, demand and price of product y_i. If the products can be substituted one for another to a greater or lesser extent, then the demand for product y_i can be written as

$$D_i = F_i(P_1, \ldots, P_N), \qquad i = 1, \ldots, N,$$

where F_i is a non-linear function with known (or estimated) coefficients such as

$$F_1 = a_0 - a_1 P_1 - a_2 P_1^2 + a_3 P_2 + a_4 P_2^2,$$

where $a_0, \ldots, a_4 > 0$. Supply is also a function of the prices, again not necessarily a linear function, so that

$$S_i = G_i(P_1, \ldots, P_N), \qquad i = 1, \ldots, N.$$

If the market is balanced, then

$$S_i = D_i, \qquad i = 1, \ldots, N$$

and the model can be represented as a system of non-linear equations

$$\mathbf{f}(\mathbf{x}) = 0,$$

where

$$\mathbf{x} \equiv [P_1, \ldots, P_N]^{\mathrm{T}} \qquad \text{and} \qquad \mathbf{f} \equiv [F_1 - G_1, \ldots, F_N - G_N]^{\mathrm{T}}.$$

The solution of this non-linear system is the system of prices to be charged to achieve a market in equilibrium.

(d) Electric power networks

Large sparse linear equations arise in the study of power networks. A power network contains a number of power stations connected by a network of transmission lines to a number of customers and the object of the analysis is to investigate the flow, subject to given load conditions.

Assume that there are M branches and N nodes. Along the jth branch of the network

$$\Delta p_j = k_j f_j, \qquad j = 1, \ldots, M$$

where, in electrical networks, Δp_j is the voltage drop, k_j is the resistance (ohms) and f_j is the current (amperes). If, at the ith node, the input is F_i, where $F_i < 0$ indicates a flow out of the system, then

$$\sum_j c_{ij} f_j = F_i, \qquad i = 1, \ldots, N$$

in which

$$c_{ij} = \begin{cases} 1 & \text{flow along branch } j \text{ towards node } i, \\ -1 & \text{flow along branch } j \text{ away from node } i, \\ 0 & \text{node } i \text{ not on branch } j. \end{cases}$$

The matrix $C = \{c_{ij}\}$ is the *incidence matrix* of the network and this latter can be interpreted as a *directed graph* (see Section 4.2). It follows that the pressure drops Δp_j can be written in terms of the nodal voltages P_i as

$$\sum_i c_{ij} P_i = \Delta p_j, \qquad j = 1, \ldots, M.$$

These three sets of equations can be combined as

$$A\mathbf{x} = \mathbf{b},$$

where

$$A = C \operatorname{diag}(k_j^{-1}) C^{\mathrm{T}},$$
$$\mathbf{x} = [P_1, \ldots, P_N]^{\mathrm{T}},$$
$$\mathbf{b} = [F_1, \ldots, F_N]^{\mathrm{T}}.$$

The same sets of equations can be used to analyse the flow in a grid of gas pipelines or any other flow network.

(e) Differential equations

In each of the preceding examples an equilibrium state was reached, but in many problems the transient state is of interest. Consider the photolysis of nitrogen dioxide, which is the first step in the production of photochemical smog:

$$NO_2 + \text{sunlight} \rightarrow NO + O \qquad (\text{reaction rate } k_1),$$
$$O + O_2 + N_2 \rightarrow O_3 + N_2 \qquad (\text{reaction rate } k_2),$$
$$O + 2O_2 \rightarrow O_3 + O_2 \qquad (\text{reaction rate } k_3),$$
$$NO + O_3 \rightarrow NO_2 + O_2 \qquad (\text{reaction rate } k_4).$$

If the concentration of O_3 is c_1, the concentration of O is c_2, the concentration of NO_2 is c_3 and the concentration of NO is c_4, then the four reactions lead to a system of coupled differential equations. If it is assumed that the supplies of sunlight, N_2 and O_2 are sufficiently large, this system can be written as

$$\begin{aligned}
\frac{dc_1}{dt} &= (k_2 + k_3)c_2 - k_4 c_1 c_4, \\
\frac{dc_2}{dt} &= k_1 c_3 - (k_2 + k_3)c_2, \\
\frac{dc_3}{dt} &= -k_1 c_3 + k_4 c_1 c_4, \\
\frac{dc_4}{dt} &= k_1 c_3 - k_4 c_1 c_4.
\end{aligned} \qquad (1.6)$$

The initial concentrations $c_1(0)$, $c_2(0)$, $c_3(0)$ and $c_4(0)$ are all known and the system can be written as

$$\frac{d\mathbf{c}}{dt} = \mathbf{F}(\mathbf{c}), \qquad \text{given } \mathbf{c}(0).$$

For some value $\Delta t > 0$, let $t_n = n\,\Delta t$ and let $\mathbf{C}_n$ be the finite-difference approximation to $\mathbf{c}(t_n)$ such that

$$\frac{1}{\Delta t}\{\mathbf{C}_{n+1} - \mathbf{C}_n\} = \tfrac{1}{2}\{\mathbf{F}(\mathbf{C}_{n+1}) + \mathbf{F}(\mathbf{C}_n)\}. \tag{1.7}$$

Proceeding from time t_n to time t_{n+1}, knowing $\mathbf{C}_n$, it is necessary to determine the $\mathbf{C}_{n+1}$ that satisfies this non-linear algebraic system. Very often the method used is a *predictor–corrector* method, which uses the formula (1.7) together with a less accurate form to provide a value for $\mathbf{C}_{n+1}$ explicitly. Such a formula is

$$\frac{1}{\Delta t}\{\mathbf{C}_{n+1} - \mathbf{C}_n\} = \mathbf{F}(\mathbf{C}_n). \tag{1.8}$$

Given $\mathbf{C}_n$, a sequence $\{\mathbf{x}^{(k)}\}$ of approximations to $\mathbf{C}_{n+1}$ is produced, where $\mathbf{x}^{(0)}$ satisfies

$$\frac{1}{\Delta t}\{\mathbf{x}^{(0)} - \mathbf{C}_n\} = \mathbf{F}(\mathbf{C}_n)$$

and where

$$\frac{1}{\Delta t}\{\mathbf{x}^{(k+1)} - \mathbf{C}_n\} = \tfrac{1}{2}\{\mathbf{F}(\mathbf{x}^{(k)}) - \mathbf{F}(\mathbf{C}_n)\}, \qquad k = 0, 1, \ldots.$$

This second equation defines an iteration of the type investigated in Chapter 5. Frequently such a simple iteration does not converge and the non-linear equation (1.7) has to be solved by alternative methods as in Section 5.8.

(f) Partial differential equations

It is probable that the largest single area in which large sparse systems of algebraic equations arises is the solution of partial differential equations. A large number of such problems involve diffusion: diffusion of heat, diffusion of pollutants in the atmosphere or in rivers, and so on. The diffusion of heat is governed by Fourier's law of heat conduction

$$\frac{\partial \theta}{\partial t} = \frac{k}{\rho c}\left(\frac{\partial^2 \theta}{\partial x^2} + \frac{\partial^2 \theta}{\partial y^2} + \frac{\partial^2 \theta}{\partial z^2}\right), \tag{1.9}$$

where

θ = temperature,

k = coefficient of thermal conductivity,

c = specific heat

and ρ = density.

The coefficient $k/\rho c$ is called the thermal diffusivity and is denoted by α. Equations similar to (1.9) can be derived for other forms of diffusion. If the derivatives in (1.9) are replaced by finite differences, then it is necessary to replace the continuous function θ by a sequence of approximate values defined at mesh points as in the preceding example. An alternative approach based on finite elements also involves a grid of nodal values. In this example each time level t_n corresponds to a three-dimensional network of points. In the resulting system of linear equations that have to be solved at each time level, the structure is similar to the power network problems. The mesh points are the nodes and the positions of the branches joining near neighbours are governed by the choice of difference formula. Unlike the power networks and the economic models, the finite-difference or finite-element model has a regular structure that can be exploited in the solution procedure. For this reason it is appropriate to devote the whole of Section 4.3 to the solution of such structured problems; in addition, Section 5.5 deals with methods frequently applied to such problems.

The construction of the difference schemes themselves is beyond the scope of this book and the reader should consult a detailed text such as Mitchell (1969) or Mitchell and Wait (1977).

PART I

Linear Algebra

2

Direct Methods for Linear Equations

2.1 Introduction

A direct method is an algorithm with a finite and *predetermined* number of steps, at the end of which it provides a solution. If the computer were capable of working with *exact arithmetic* the solution thus obtained would be exact. Unfortunately computers have a finite word length and therefore all numbers output as part of the 'solution' are subject to round-off errors. In some problems this may mean that the computed solution is worthless. It is therefore of paramount importance that the computer user should be fully aware of the limitations of the algorithms being used – preferably before using a numerical method, but definitely before using the results.

The basis of most tried and tested methods is that the solution of the system $A\mathbf{x} = \mathbf{b}$ is unaltered if:

(1) the rows of A and $\mathbf{b}$ are reordered
(2) a row is modified by adding an arbitrary multiple of any other row.

Thus it is possible to transform the system $A\mathbf{x} = \mathbf{b}$ into a system $U\mathbf{x} = \mathbf{y}$ where U is an upper triangular matrix, such that the two systems have the same solution. The solution of $U\mathbf{x} = \mathbf{y}$ can then be found by *back substitution.* The operations of interchanging rows of a matrix and of combining rows can be represented as *elementary transformations* of the original matrix. Thus, for example, interchanging the second and fifth rows of a 5×5 matrix A is equivalent to premultiplying A by a *permutation matrix P*, where in this particular example

$$P = \begin{bmatrix} 1 & 0 & 0 & 0 & 0 \\ 0 & 0 & 0 & 0 & 1 \\ 0 & 0 & 1 & 0 & 0 \\ 0 & 0 & 0 & 1 & 0 \\ 0 & 1 & 0 & 0 & 0 \end{bmatrix}.$$

In general a permutation matrix is a matrix of zeros and ones with exactly one non-zero component in each row and column. Similarly if γ times the second row is to be added to the fifth row, the equivalent transformation involves premultiplying by

$$L = \begin{bmatrix} 1 & & & & \\ 0 & 1 & & & \\ 0 & 0 & 1 & & \\ 0 & 0 & 0 & 1 & \\ 0 & \gamma & 0 & 0 & 1 \end{bmatrix}.$$

Exercises

2.1. Verify that the simultaneous modification of rows $(j + 1), \ldots, N$ of an $N \times N$ matrix, by adding to them respectively, $\gamma_{j+1}, \ldots, \gamma_N$ times the jth row, is equivalent to premultiplying by the $N \times N$ transformation matrix

$$L = \begin{bmatrix} 1 & & & & \\ & \ddots & & & \\ & & 1 & & \\ & & \gamma_{j+1} & 1 & \\ & & \vdots & & \ddots \\ & & \gamma_N & & & 1 \end{bmatrix}.$$
$$\qquad\qquad j$$

2.2. Verify that modification of a number of rows by any other row is equivalent to a transformation of the form PLP, where L is *lower triangular* and P is a permutation matrix.

2.2 Gauss Elimination

The Gauss elimination algorithm generates a sequence of matrices $\{A^{(k)}\}$ $(k = 1, \ldots, N)$, where $A^{(1)} \equiv A$ and a sequence of right-hand sides $\{\mathbf{b}^{(k)}\}$ $(k = 1, \ldots, N)$ such that $A^{(N)} \equiv U$ is upper triangular and such that *if exact arithmetic is used for the transformations*, then the linear systems

$$A^{(k)}\mathbf{x} = \mathbf{b}^{(k)}, \qquad k = 1, \ldots, N$$

have the same solution. In practice round-off errors will enter the calculation and (see Wilkinson 1963) the solution of

$$U\mathbf{x} = \mathbf{y}$$

is in fact the solution of a *perturbed system*. That is a system of the form

$$(A + \delta A)\mathbf{x} = \mathbf{b} + \delta \mathbf{b}. \tag{2.1}$$

It is to be hoped that the matrix δA and the vector $\delta \mathbf{b}$ are small in relation to A and $\mathbf{b}$, so that the solution of (2.1) is not too dissimilar from the solution of $A\mathbf{x} = \mathbf{b}$.

The first derived system $A^{(2)}\mathbf{x} = \mathbf{b}^{(2)}$ is obtained by eliminating the *subdiagonal components* in the first column by subtracting suitable multiples of the first row from each of the subsequent rows. Thus

$$A^{(2)} = \begin{bmatrix} a_{11}^{(1)} & a_{12}^{(1)} & \dots & a_{1N}^{(1)} \\ 0 & a_{22}^{(2)} & \dots & a_{2N}^{(2)} \\ \vdots & \vdots & & \vdots \\ 0 & a_{N2}^{(2)} & \dots & a_{NN}^{(2)} \end{bmatrix},$$

where, assuming $a_{11}^{(1)} \neq 0$,

$$a_{ij}^{(2)} = a_{ij}^{(1)} - m_{i1} a_{1j}^{(1)}, \qquad i, j = 2, \dots, N$$

with

$$m_{i1} = \frac{a_{i1}^{(1)}}{a_{11}^{(1)}}.$$

If the modification formula is rewritten as

$$a_{ij}^{(1)} = a_{ij}^{(2)} + m_{i1} a_{1j}^{(1)},$$

it is clear that as the first row $\{a_{1j}^{(1)}\}$ is unaltered, it follows that the matrices $A^{(1)}$ and $A^{(2)}$ satisfy

$$A^{(1)} = L^{(1)} A^{(2)}$$

where

$$L^{(1)} = \begin{bmatrix} 1 & & & \\ m_{21} & 1 & & \\ \vdots & & \ddots & \\ m_{N1} & & & 1 \end{bmatrix}.$$

Similarly if

$$a_{22}^{(2)} = a_{22}^{(1)} - m_{21} a_{12}^{(1)} \neq 0,$$

i.e. if

$$\det\begin{bmatrix} a_{11}^{(1)} & a_{12}^{(1)} \\ a_{21}^{(1)} & a_{22}^{(1)} \end{bmatrix} \neq 0,$$

it follows that the formulae

$$a_{ij}^{(3)} = a_{ij}^{(2)} - m_{i2}a_{2j}^{(2)}, \qquad i, j = 3, \ldots, N$$

and

$$m_{i2} = a_{i2}^{(2)}/a_{22}^{(2)}$$

define a matrix $A^{(3)}$ with zero subdiagonal components in the second column, assuming that the first and second rows remain unchanged.

In general for $k = 1, 2, \ldots, N-1$, if $a_{kk}^{(k)} \neq 0$

$$\left.\begin{aligned} a_{ij}^{(k+1)} &= a_{ij}^{(k)} - m_{ik}a_{kj}^{(k)}, \qquad i, j = k+1, \ldots, N, \\ \text{with} \\ m_{ik} &= \frac{a_{ik}^{(k)}}{a_{kk}^{(k)}}, \end{aligned}\right\} \tag{2.2}$$

then

$$A^{(k)} = L^{(k)}A^{(k+1)},$$

where

$$L^{(k)} = \begin{bmatrix} 1 & & & & & \\ & \ddots & & & & \\ & & 1 & & & \\ & & m_{k+1,k} & 1 & & \\ & & \vdots & & \ddots & \\ & & m_{Nk} & & & 1 \end{bmatrix}.$$

The right-hand sides can be similarly transformed into

$$b_i^{(k+1)} = b_i^{(k)} - m_{ik}b_k^{(k)},$$

and it follows that

$$\mathbf{b}^{(k)} = L^{(k)}\mathbf{b}^{(k+1)}, \qquad k = 1, \ldots, N-1.$$

Thus the constituents of the triangular system $A^{(N)}\mathbf{x} = \mathbf{b}^{(N)}$, where

$$A^{(N)} = \begin{bmatrix} a_{11}^{(1)} & a_{12}^{(1)} & \ldots & a_{1N}^{(1)} \\ & a_{22}^{(2)} & \ldots & a_{2N}^{(2)} \\ & & \ddots & \vdots \\ & & & a_{NN}^{(N)} \end{bmatrix} \text{ and } \mathbf{b}^{(N)} = \begin{bmatrix} b_1^{(1)} \\ b_2^{(2)} \\ \vdots \\ b_N^{(N)} \end{bmatrix}$$

tisfy

$$A = L^{(1)} \dots L^{(N-1)} A^{(N)}$$

d

$$\mathbf{b} = L^{(1)} \dots L^{(N-1)} \mathbf{b}^{(N)}.$$

ritten in this form we say that we have *factorized* the matrix A. The method as rmulated in (2.2) breaks down if at any stage $a_{kk}^{(k)} = 0$. Even when the values of $a_{kk}^{(k)}$ (known as *pivots*) are not zero, round-off errors can ruin the accuracy of the ial solution if they become very small.

It is vital to remember that Gauss elimination should never by used in this simple rm (2.2), but that *it always involves some pivoting strategy to reduce the effects round-off errors.*

ack substitution

Once the matrix has been reduced to upper-triangular form, it is possible to solve r the components of $\mathbf{x}$ in the reverse order using the kth equation to solve for the h component x_k $(k = N, \dots, 1)$. Thus, deleting superscripts for clarity it follows at

$$x_N = b_N / a_{NN}$$

d then

$$x_k = (b_k - \sum_{j=k+1}^{N} a_{kj} x_j) / a_{kk}, \qquad k = N-1, \dots, 1.$$

Exercise

3. Let $\mathbf{x} = \mathbf{x}_0$ be the solution of the linear system of equations

$$A\mathbf{x} = \mathbf{b}$$

d $\mathbf{x} = \mathbf{x}_0 + \delta\mathbf{x}$ be the solution of the perturbed system of equations

$$(A + \delta A)\mathbf{x} = \mathbf{b} + \delta\mathbf{b}.$$

ow that the relative error $\dfrac{\|\delta\mathbf{x}\|}{\|\mathbf{x}_0\|}$ is bounded by

$$\frac{\|\delta\mathbf{x}\|}{\|\mathbf{x}_0\|} \leqslant \frac{\|A^{-1}\|\,\|A\|}{1 - \|A^{-1}\|\,\|\delta A\|} \left\{ \frac{\|\delta\mathbf{b}\|}{\|\mathbf{b}\|} + \frac{\|\delta A\|}{\|A\|} \right\}$$

$\|A^{-1}\|\,\|\delta A\| < 1$. It is assumed here and in later chapters that the reader is miliar with matrix and vector norms.

2.3 Pivoting Strategy

A simple pivoting strategy that is sufficient in most cases is *partial pivoting*; this volves the interchange of the pth and kth rows $(p > k)$ of the matrix $A^{(k)}$, where

p is chosen such that

$$|a_{pk}^{(k)}| = \max_{k \leqslant i \leqslant N} |a_{ik}^{(k)}|,$$

thus it follows that

$$|m_{ik}| \equiv \left|\frac{a_{ik}}{a_{pk}}\right| \leqslant 1.$$

When the elimination is being performed automatically it is often quicker to store the order of the pivot rows as a vector of indices $[p_1, \ldots, p_N]^T$ rather tha physically interchange positions of the rows in the computer.

When partial pivoting is used, Gauss elimination is equivalent to a transformati of the form

$$PA = L^{(1)} \cdots L^{(N-1)} A^{(N)}$$

where P is a permutation matrix.

Not only does partial pivoting keep the effects of round-off error down to an acceptable level in most problems, but for a non-singular matrix it will never brea down due to a zero pivot (excluding the effect of round-off errors in almost singular problems). If at any stage the method failed to find a non-zero pivot it would mean that after the interchanges had been performed there was some k suc that:

$$A^{(k)} = \begin{bmatrix} a_{11}^{(1)} & & & & a_{1N}^{(1)} \\ & \ddots & & & \vdots \\ & & a_{kk}^{(k)} & a_{kk} \quad \cdots & a_{kN}^{(k)} \\ & & 0 & & \\ & & \vdots & C^{(k)} & \\ & & 0 & & \end{bmatrix},$$

where the $(N-k) \times (N-k)$ submatrix $C^{(k)}$ is of the form

$$C^{(k)} = \begin{bmatrix} 0 & a_{k+1,k+2}^{(k)} & \cdots & a_{k+1,N}^{(k)} \\ \vdots & & & \vdots \\ 0 & a_{N,k+2}^{(k)} & \cdots & a_{NN}^{(k)} \end{bmatrix}.$$

Clearly $C^{(k)}$ is *singular*, and so therefore is $A^{(k)}$, and hence A is singular.

Cautionary example (The necessity for a sensible pivoting strategy) Given the trix

$$A = \begin{bmatrix} 4 & 3 & 2 & 1 \\ 4+\epsilon & 3 & 5-\epsilon & 4 \\ 4+\epsilon & 3+\epsilon & 1 & 2 \\ 5 & 2 & -1 & 1 \end{bmatrix}$$

d the vector $\mathbf{b} = [10, 16, 10 + 2\epsilon, 7]^T$, the solution of the system $A\mathbf{x} = \mathbf{b}$ is $= [1, 1, 1, 1]^T$ for any value of the small parameter ϵ. If, however, the system is ved numerically for $\epsilon = 10^{-5}$, using Gauss elimination *without pivoting*, the ults are somewhat different. In the matrices $A^{(k)}$, an asterisk (*) indicates the ue is unaltered and the values below the diagonal in parentheses indicate a zero mponent of $A^{(k)}$ is overwritten by the corresponding value of m_{ik} in the factor $^{k)}$. In order to simplify the presentation, very large numbers and very small mbers are included in a floating point format, thus

$$-7.499094 \text{ E} - 6 \equiv -7.499094 \times 10^{-6} \equiv -0.000007499094.$$

$$A^{(2)} = \begin{bmatrix} 4 & 3 & 2 & 1 \\ (1.0000025) & -7.4990094 \text{ E} - 6 & 2.9999850 & 2.9999975 \\ (1.0000025) & 2.4996668 \text{ E} - 6 & -1.0000050 & 0.9999975 \\ (1.25) & -1.75 & -3.5 & -0.25 \end{bmatrix},$$

$$\mathbf{b}^{(2)} = [10, \quad 5.9999750, \quad -5.0067902 \text{ E} - 6, \quad -5.5]^T,$$

$$A^{(3)} = \begin{bmatrix} * & * & * & * \\ * & * & * & * \\ * & (-0.3333333) & -9.9986791 \text{ E} - 6 & 1.9999967 \\ * & (2.3336416 \text{ E } 5) & -7.0000247 \text{ E } 5 & -7.0009214 \text{ E } 5 \end{bmatrix},$$

$$\mathbf{b}^{(3)} = [10, \quad 5.9999750, \quad 1.9999867, \quad -1.4001846 \text{ E } 6]^T,$$

$$A^{(4)} = \begin{bmatrix} * & * & * & * \\ * & * & * & * \\ * & * & * & * \\ * & * & (7.0018495 \text{ E } 10) & -1.4003746 \text{ E } 11 \end{bmatrix},$$

$$\mathbf{b}^{(4)} = [10, \quad 5.9999750, \quad 1.9999867, \quad -1.4003746 \text{ E } 11]^T.$$

ck substitution leads to the solution

$$\mathbf{x} = [168.7, \quad -222.6, \quad 0.9994, \quad 1]^T!$$

Reordering the equations in the reverse order leads to

$$A^{(2)} = \begin{bmatrix} 5 & 2 & -1 & 1 \\ (0.800002) & 1.400006 & 1.800002 & 1.199998 \\ (0.800002) & 1.399996 & 5.799992 & 3.199998 \\ (0.8) & 1.4 & 2.8 & 0.2 \end{bmatrix},$$

$$\mathbf{b}^{(2)} = \begin{bmatrix} 7 \\ 4.4 \\ 10.399986 \\ 4.4 \end{bmatrix},$$

$$A^{(3)} = \begin{bmatrix} * & * & * & * \\ * & * & * & * \\ * & (0.9999929) & 4.0000029 & 2.0000086 \\ * & (0.9999957) & 1.0000057 & 0.9999929 \end{bmatrix},$$

$$\mathbf{b}^{(3)} = [7, \quad 4.4, \quad 6.0000114, \quad 1.2852251 \text{ E} - 5]^{\mathrm{T}},$$

$$A^{(4)} = \begin{bmatrix} * & * & * & * \\ * & * & * & * \\ * & * & * & * \\ * & * & (0.2500012) & -1.4999975) \end{bmatrix},$$

$$\mathbf{b}^{(4)} = [7, \quad 4.4, \quad 6.0000114, \quad -1.4999975]^{\mathrm{T}}.$$

Back substitution leads to the solution

$$\mathbf{x} = [1, 1, 1, 1]^{\mathrm{T}}.$$

Complete pivoting

An alternative strategy known as *complete pivoting* is to perform *row and column interchanges*; this implies that the solution vector will have been reordered. At the kth stage the pivot is chosen as $a_{pq}^{(k)}$ such that

$$|a_{pq}^{(k)}| = \max_{k \leqslant i,j \leqslant N} |a_{ij}^{(k)}| \equiv \max\{|a_{ij}^{(k)}| : a_{ij}^{(k)} \in C^{(k)}\}.$$

Complete pivoting is suggested as a means of reducing the *growth* of components of U. If only partial pivoting is used, then as $|m_{ik}| \leqslant 1$ it follows from (2.2) that

$$|a_{ij}^{(k+1)}| \leqslant |a_{ij}^{(k)}| + |a_{kj}^{(k)}| \tag{2..}$$

and so it is possible that

$$|a_{ij}^{(k+1)}| \cong 2\,|a_{ij}^{(k)}|.$$

It follows from (2.3) that

$$\begin{aligned}\max_{i,j} |a_{ij}^{(k+1)}| &\leqslant 2 \max_{i,j} |a_{ij}^{(k)}| \\ &\leqslant 2^{k-1} \max_{i,j} |a_{ij}|.\end{aligned}$$

By using an example (originally due to Wilkinson 1961), it is possible to show that this doubling can occur (see Exercise 2.4)

Exercises

2.4. Reduce the matrix

$$A = \begin{bmatrix} 1 & & & & & 1 \\ -1 & 1 & & & & 1 \\ -1 & -1 & 1 & & & 1 \\ -1 & -1 & -1 & 1 & & 1 \\ -1 & -1 & -1 & -1 & 1 & 1 \\ -1 & -1 & -1 & -1 & -1 & 1 \end{bmatrix}$$

to triangular form using (a) partial pivoting, (b) complete pivoting.

2.5. Verify that Gaussian elimination, when applied with complete pivoting, is equivalent to the transformation

$$P_{\mathrm{r}} A P_{\mathrm{c}} = L^{(1)} \cdots L^{(N-1)} A^{(N)},$$

where P_{r} and P_{c} are permutation matrices that give the reordering of the rows and columns respectively.

2.6. Verify that $L \equiv L^{(1)} \cdots L^{(N-1)}$ is a lower-triangular matrix.

2.7. Prove that if the matrix A is diagonally dominant *by columns*, i.e.

$$\sum_{\substack{i=1 \\ i \neq j}}^{N} |a_{ij}| \leqslant |a_{jj}|, \qquad j = 1, \ldots, N,$$

then the matrices $C^{(k)}$ $(k = 1, \ldots, N-1)$ are all diagonally dominant by columns. Hence verify that partial pivoting does not alter the order of the rows of such a matrix.

2.4 Triangular Factorization

It is possible to use Gauss elimination with partial pivoting to provide a *constructive proof* of a basic theorem of matrix algebra.

Theorem *For any non-singular matrix A, there exists at least one permutation matrix P such that the matrix PA can be uniquely factorized as*

$$PA = LDU,$$

where L is a lower-triangular matrix with a unit diagonal, U is an upper-triangular matrix with a unit diagonal, and D is a diagonal matrix. Given P such that $PA = LD$ holds, then L, D and U are unique. Δ

The symbol Δ is used throughout this book to denote the logical end of both th statement of a theorem and the proof.

The diagonal matrix may be imbedded in either L by scaling the columns, or U by scaling the rows, if components other than unity are allowed on the diagonal. The resulting factorizations are known as the *Crout factorization* and the *Doolittle factorization* respectively.

Proof of Existence It follows from the preceding statement of Gauss elimination with partial pivoting that at least one such permutation matrix exists such that

$$PA = L^{(1)} \cdots L^{(N-1)}A^{(N)}.$$

The matrix $A^{(N)}$ is upper triangular and the elementary matrices $L^{(1)}, \ldots, L^{(N-1)}$ are such that $L \equiv L^{(1)} \cdots L^{(N-1)}$ is lower triangular and has a unit diagonal. Since the diagonal components of $A^{(N)}$ are non-zero it follows that

$$D = \text{diag}\,(a_{kk}^{(k)})$$

and

$$U = D^{-1}A^{(N)}. \quad \Delta$$

Proof of Uniqueness Assume triangular factors of the matrix A exist; then the matrix equation

$$A = LDU$$

can be written in component form as

$$a_{ij} = \sum_{k=1}^{\min(i,j)} l_{ik}d_{kk}u_{kj}. \tag{2.4}$$

If it is assumed that *U has a unit diagonal*, then $j = 1$ in (2.4) leads to

$$\left.\begin{aligned} a_{11} &= d_{11}, \\ a_{i1} &= l_{i1}d_{11}, \qquad i = 2, \ldots, N \end{aligned}\right\}, \tag{2.5a}$$

and $i = 1$ leads to

$$a_{1j} = d_{11}u_{1j}, \qquad j = 2, \ldots, N. \tag{2.5b}$$

Thus the first row and column of L, D, and U are uniquely defined. The proof then proceeds inductively: assume that the first $i - 1$ rows and columns of L, D, and U are defined uniquely; then equating coefficients in the remainder of the ith column gives

$$a_{ji} = \sum_{k=1}^{i} l_{jk} d_{kk} u_{kj}, \qquad j = i, \ldots, N,$$

but $l_{jj} = u_{ii} = 1$, so

$$a_{ii} = d_{ii} + \sum_{k=1}^{i-1} l_{jk} d_{kk} u_{ki}$$

and (2.5c)

$$a_{ji} = l_{ji} d_{ii} + \sum_{k=1}^{i-1} l_{jk} d_{kk} u_{ki}, \qquad j = i+1, \ldots, N.$$

As all the terms in the summations are known, d_{ii} and the l_{ji} ($j = i + 1, \ldots, N$) are uniquely defined. Similarly, in the remainder of the ith row

$$a_{ij} = \sum_{k=1}^{i} l_{ik} d_{kk} u_{kj}$$

$$= d_{ii} u_{ij} + \sum_{k=1}^{i-1} l_{ik} d_{kk} u_{kj}, \qquad j = i+1, \ldots, N, \qquad (2.5d)$$

where all the terms apart from u_{ij} have been previously defined. Since the factorization is assumed to exist, $d_{ii} \neq 0$ and thus the u_{ij}($j = i + 1, \ldots, N$) are uniquely determined. Δ

Crout's method

The constructive proof of uniqueness leads directly to an algorithm for computing *LDU* factors. However, it is often preferable to compute the *LU* factors from (2.5a)–(2.5d), with $l_{ij} d_{jj}$ replaced by l_{ij}, providing that in a practical algorithm, at least partial pivoting is employed. That means for each i, if

$$| l_{pi} | = \max_{i \leqslant k \leqslant N} | l_{ki} |,$$

then the pth and ith rows will be interchanged. With this proviso, the first steps of the algorithm can be written as

$$l_{i1} = a_{i1}, \qquad j = 1, \ldots, N,$$

and

$$u_{1j} = \frac{a_{1j}}{l_{11}}, \qquad j = 2, \ldots, N.$$

Then for each $i = 2, \ldots, N$

$$l_{ji} = a_{ji} - \sum_{k=1}^{i-1} l_{jk} u_{ki}, \qquad j = i, \ldots, N,$$

and

$$u_{ij} = \frac{a_{ij} - \sum_{k=1}^{i-1} l_{ik} u_{kj}}{l_{ii}}, \qquad j = i + 1, \ldots, N.$$

This form has certain practical advantages over the Gauss elimination algorithm, particularly if multiple precision is available or if equations of the form $A\mathbf{x} = \mathbf{b}_k$ are to be solved for a number of different right-hand sides $\mathbf{b}_k$ $(k = 1, \ldots, M)$. Once the factorization is complete it only remains to solve sequentially the triangular systems

$$L\mathbf{y} = \mathbf{b}$$

and

$$U\mathbf{x} = \mathbf{y},$$

to obtain the solution of $A\mathbf{x} = \mathbf{b}$. This is a two-stage process consisting of

(1) *Forward substitution*

$$y_1 = \frac{b_1}{l_{11}},$$

$$y_k = \frac{b_k - \sum_{j=1}^{k-1} l_{kj} y_j}{l_{kk}}, \qquad k = 2, \ldots, N.$$

(2) *Backward substitution*

$$x_n = y_n,$$

$$x_k = y_k - \sum_{j=k+1}^{N} u_{kj} x_j, \qquad k = N - 1, \ldots, 1.$$

Triangular factorization involves significantly more work in terms of arithmetic operations than either form of substitution (see Exercise 2.8).

Numerical example (Crout's method using partial pivoting) As in the earlier example

$$A = \begin{bmatrix} 4 & 3 & 2 & 1 \\ 4+\epsilon & 3 & 5-\epsilon & 4 \\ 4+\epsilon & 3+\epsilon & 1 & 2 \\ 5 & 2 & -1 & 1 \end{bmatrix} \text{ and } \mathbf{b} = \begin{bmatrix} 10 \\ 16 \\ 10+2\epsilon \\ 7 \end{bmatrix},$$

with $\epsilon = 10^{-5}$, then Crout's method with row interchanges leads to reversal of the order of the rows of A and $\mathbf{b}$, the same as Gauss elimination with row interchanges. The factors of the reordered matrix were computed as

$$L = \begin{bmatrix} 5 & & & \\ 4.00001 & 1.4000060 & & \\ 4.00001 & 1.3999960 & 4.0000029 & \\ 4 & 1.4000000 & 1.0000057 & -1.4999975 \end{bmatrix}$$

and

$$U = \begin{bmatrix} 1 & 0.4000000 & -0.2000000 & 0.2000000 \\ & 1 & 1.2857102 & 0.8571378 \\ & & 1 & 0.5000000 \\ & & & 1 \end{bmatrix}.$$

The solution of $L\mathbf{y} = \mathbf{b}^*$, where $\mathbf{b}^*$ is the reordered right-hand side, was

$$\mathbf{y} = [1.4000000, 3.1428480, 1.5000018, 1.0000000]^{\mathrm{T}};$$

finally the solution of $U\mathbf{x} = \mathbf{y}$ was found to be $\mathbf{x} = [1, 1, 1, 1]^{\mathrm{T}}$. Note that the columns of L correspond to columns of $A^{(4)}$ and rows of U are the rows of $A^{(4)}$ are suitable scaling.

Lemma *If the matrix A is symmetric and positive definite*, then there exists a unique lower-triangular matrix L with positive diagonal such that*

$$A = LL^{\mathrm{T}}. \Delta$$

This is known as the *Cholesky factorization.* It is widely used in practice, but it is often preferable even for positive definite matrices to use the factorization

$$A = LDL^{\mathrm{T}},$$

where L has a unit diagonal (see for example Gill and Murray, 1972).

Exercise

2.8. Verify that both triangular factorization and Gauss elimination require $\frac{1}{3}N^3 + O(N^2)$ arithmetic operations, whereas the substitution (backward or forward) requires only $O(N^2)$ arithmetic operations.

Matrix scaling

An important feature of an efficient numerical algorithm for solving linear equations is determining a suitable scaling of the equations. The control of the

*Positive definite means that for some $\gamma > 0$, $\mathbf{x}^{\mathrm{T}}A\mathbf{x} \geqslant \gamma\mathbf{x}^{\mathrm{T}}\mathbf{x}$ for all $\mathbf{x}$.

round-off errors in an elimination is invariably enhanced if the rows of A and $\mathbf{b}$ are properly scaled, and a sensible choice is to ensure that the coefficients are as nearly as possible of the same order of magnitude. Scaling of the rows clearly leaves the solution unaltered, and scaling of the columns involves a simple scaling of components of the solution. So it is possible to use scaling of rows and columns before proceeding for elimination (see Curtis and Reid, 1972). On computers using binary arithmetic, any *scaling should be by powers of 2*, since this does not alter the accuracy of the coefficients.

Iterative refinement

A rigorous analysis of the effects of round-off errors when solving linear systems shows that the computed solution can be written as the exact solution of a perturbed system (2.1). Any attempt to estimate the size of the error in the solution involves an estimate of the condition number

$$K(A) = \| A^{-1} \| \, \| A \|$$

(see Exercise 2.3). As the condition number is not usually available, other methods of estimating the error have to be used.

One popular and very satisfactory practical method is *iterative refinement*: Using the computed solution $\bar{\mathbf{x}}$ evaluate the *residual vector*

$$\mathbf{r}(\bar{\mathbf{x}}) \equiv \mathbf{b} - A\bar{\mathbf{x}},$$

using double precision if the results are to be meaningful. Then using the previously computed triangular factors of A, solve the equations

$$A\,\delta\mathbf{x} = \mathbf{r}. \tag{2.6}$$

If $\| \delta\mathbf{x} \| / \| \bar{\mathbf{x}} \|$ is 'small' then the original result is satisfactory, if it is not, then the computed solution is replaced by $\bar{\mathbf{x}} + \delta\mathbf{x}$ and the refinement procedure is repeated using the residual $\mathbf{r}(\bar{\mathbf{x}} + \delta\mathbf{x})$ (computed using double precision) in the right-hand side of (2.6).

For well-conditioned equations with

$$K(A)2^{-t} < 1, \tag{2.7}$$

where it is assumed the computer retains t binary digits, this iterative refinement should eventually converge to roughly single-precision accuracy. If (2.7) is not satisfied it will soon be apparent that the refinement is not converging and the last correction $\delta\mathbf{x}$ is taken as an indication of the size of the errors.

One side effect of iterative improvement is that it does provide a reasonable estimate of the condition number of the matrix A. It follows from Exercise 2.3 that when $\| \delta A \| = 0$,

$$\frac{\| \delta\mathbf{x} \|}{\| \mathbf{x} \|} \leqslant K(A) \frac{\| \delta\mathbf{b} \|}{\| \mathbf{b} \|}.$$

This initial assumption shows that we intend to estimate the condition number of the matrix $\bar{A} = LU$ where L and U are the *computed* factors. Then

$$\delta \mathbf{x} = \bar{\mathbf{x}} - \bar{A}^{-1}\mathbf{b},$$

and

$$\delta \mathbf{b} = A\bar{\mathbf{x}} - \mathbf{b} = -\mathbf{r}(\bar{\mathbf{x}})$$

and so

$$\delta \mathbf{x} = -\bar{A}^{-1}\mathbf{r}(\bar{\mathbf{x}}). \tag{2.8}$$

Using Gauss elimination it follows (see Forsythe and Moler, 1967) that

$$r_1 \simeq 2^{-t}$$

and

$$\| \mathbf{r} \| \simeq 2^{-t} \| \bar{A} \| \| \bar{\mathbf{x}} \|, \tag{2.9}$$

combining (2.8) and (2.9) shows that $\| \delta \mathbf{x} \|$ is an approximate lower bound on

$$2^{-t} K(A) \| \bar{\mathbf{x}} \| .$$

Thus if $\delta \mathbf{x}$ is replaced by the computed correction $\delta \bar{\mathbf{x}}$ we obtain

$$2^{t} \frac{\| \delta \bar{\mathbf{x}} \|}{\| \bar{\mathbf{x}} \|}$$

as an approximation to the condition number $K(A)$.

3

Orthogonal Factorization

3.1 Existence and Uniqueness

In Section 2.4 it is shown that a non-singular matrix can (after permuting rows if necessary) be factorized as the product of two matrices, one upper triangular and one lower triangular. Although triangular factorization is basic to various numerical methods it does have limitations, particularly with regard to numerical stability. There is an alternative form of factorization which, while involving more computation in terms of arithmetic operations, often leads to a more stable algorithm which is preferred in certain circumstances – such as in least-squares problems (Section 3.3) or factor modification (required in Section 7.3).

The alternative form is the QR or orthogonal factorization, that is a matrix A is factorized as the product of an orthogonal matrix* Q and an upper-triangular matrix R. It is occasionally found convenient to use LQ factorization (L is lower triangular) in numerical algorithms, as in Section 7.3.

It is possible to prove the existence and uniqueness of the QR factorization using a constructive proof given by Wilkinson (1965b).

Theorem *For any non-singular real-valued matrix A, there exists a unique QR factorization such that the diagonal components of R are positive.* Δ

Note The proof is both constructive and inductive but unlike the proof of the theorem on triangular factors, the particular construction used in the proof is not recommended as a suitable basis for a practical numerical method of factorizing a matrix. An efficient numerical method is based on elementary transformations such as those attributed to Givens and described in the next section. An alternative is to use Householder's transformations as described in Gourlay and Watson (1973).

*An orthogonal matrix Q is such that $Q^{\mathrm{T}} \equiv Q^{-1}$.

Proof The factors of A are constructed column by column. Thus for $s = 1, \ldots,$ N, let A_s be the $N \times s$ matrix consisting of the first s columns of A. Similarly define

$$Q_s \equiv \begin{bmatrix} q_{11} & \cdots & q_{1s} \\ \vdots & & \vdots \\ q_{N1} & \cdots & q_{Ns} \end{bmatrix}, \qquad s = 1, \ldots, N$$

to be a matrix with columns $\mathbf{q}_j = [q_{1j}, \ldots, q_{Nj}]^{\mathrm{T}}$ that are *orthonormal*, i.e.

$$\mathbf{q}_i^{\mathrm{T}}\mathbf{q}_j = \begin{cases} 1, & i = j \\ 0, & i \neq j \end{cases} \qquad 1 \leqslant i, j \leqslant s$$

and define

$$R_s \equiv \begin{bmatrix} r_{11} & \cdots & r_{1s} \\ & \ddots & \vdots \\ & & r_{ss} \end{bmatrix}, \qquad s = 1, \ldots, N,$$

such that $r_{ii} > 0$ $(i = 1, \ldots, s)$.

It is then assumed that the first $k - 1$ columns $(1 < k \leqslant N)$ have been factorized and that

$$A_{k-1} = Q_{k-1}R_{k-1}. \tag{3.1}$$

Then it is possible to define

$$\mathbf{q}_k \equiv [q_{1k}, \ldots, q_{Nk}]^{\mathrm{T}} \qquad \text{and} \qquad \mathbf{r}_k \equiv [r_{1k}, \ldots, r_{kk}]^{\mathrm{T}}$$

such that

$$A_k = Q_kR_k. \tag{3.2}$$

The only difference between (3.1) and (3.2) is the addition of one extra column. If $\mathbf{a}_k \equiv [a_{1k}, \ldots, a_{Nk}]^{\mathrm{T}}$, this additional column can be written as

$$\mathbf{a}_k = Q_k\mathbf{r}_k. \tag{3.3}$$

Since the columns of Q_k are orthonormal,

$$Q_k^{\mathrm{T}}Q_k = I_k,$$

where I_k is the $k \times k$ identity matrix; thus (3.3) can be written as

$$\mathbf{r}_k = Q_k^{\mathrm{T}}\mathbf{a}_k. \tag{3.4}$$

Since the first $k - 1$ rows of the right-hand side are equivalent to $Q_{k-1}^{\mathrm{T}}\mathbf{a}_k$, which is known, it follows from (3.4) that

$$r_{jk} = \mathbf{q}_j^{\mathrm{T}}\mathbf{a}_k, \qquad j = 1, \ldots, k - 1.$$

Thus *above the diagonal* the r_{jk} exist and are unique.

The kth column of (3.3) can be rewritten as

$$q_{ik}r_{kk} = a_{ik} - \sum_{j=1}^{k-1} q_{ij}r_{jk}, \qquad i = 1, \ldots, N, \tag{3.5}$$

and the right-hand side is known and is denoted hereafter by y_i. Since Q is to be orthogonal

$$\mathbf{q}_k^{\mathrm{T}}\mathbf{q}_k = 1$$

and hence it follows from (3.5) that

$$r_{kk}^2 = \sum_{i=1}^{N} y_i^2.$$

Thus $q_{ik} (i = 1, \ldots, N)$ are uniquely defined by (3.5) if $r_{kk} \neq 0$, but $r_{kk} = 0$ only if $y_i = 0$ $(i = 1, \ldots, k)$, that is only if

$$\mathbf{a}_k - \sum_{j=1}^{k-1} r_{jk}\mathbf{q}_j = 0. \tag{3.6}$$

It follows from (3.6) that $\mathbf{q}_1, \ldots, \mathbf{q}_{k-1}, \mathbf{a}_k$ are linearly dependent and hence (see Exercise 3.2) that $\mathbf{a}_1, \ldots, \mathbf{a}_{k-1}, \mathbf{a}_k$ are linearly dependent; thus A is singular. Since A was assumed to be non-singular, it follows that $r_{kk} \neq 0$ and it is possible tc take $r_{kk} > 0$ and define

$$q_{ik} = \frac{y_i}{r_{kk}}, \qquad i = 1, \ldots, N.$$

It only remains to note that the first column of the factorization can be written as

$$\mathbf{a}_1 = r_{11}\mathbf{q}_1,$$

thus

$$r_{11} = \left\{ \sum_{j=1}^{N} a_{j1}^2 \right\}^{1/2} > 0,$$

$$q_{j1} = \frac{a_{j1}}{r_{11}}, \qquad j = 1, \ldots, N,$$

and the proof is complete. Δ

Definition The rank of an $M \times N$ matrix is defined as:

(1) the number of linearly independent rows (or columns);
(2) the dimension of the space spanned by the rows (or columns); or
(3) the value r such that there exists a non-zero minor of order r, but no non-zero minor of order $\geqslant r + 1$.

All three definitions are equivalent.

Corollary *There exists a unique QR factorization of a real $M \times N$ $(M > N)$ matrix of rank N, such that R is an $N \times N$ matrix with positive diagonal components and such that the columns of the $M \times N$ matrix Q are orthonormal.* Δ

Exercises

3.1. Prove that the LQ factorization of a real non-singular matrix is also unique if L is chosen with positive diagonal components.

3.2. Prove that there exist constants α_{ij} $(\alpha_{jj} \neq 0)$ and β_{ij} $(\beta_{jj} \neq 0)$ such that

$$\mathbf{a}_j = \sum_{i=1}^{j} \alpha_{ij} \mathbf{q}_i \qquad \text{and} \qquad \mathbf{q}_j = \sum_{i=1}^{j} \beta_{ij} \mathbf{a}_i, \qquad j = 1, \ldots, N,$$

and hence verify that the equivalence of the following statements for any s such that $1 \leqslant s \leqslant N$:

(1) $\mathbf{a}_1, \ldots, \mathbf{a}_s$ are linearly dependent,
(2) $\mathbf{q}_1, \ldots, \mathbf{q}_s$ are linearly dependent,
(3) $\mathbf{q}_1, \ldots, \mathbf{q}_j, \mathbf{a}_{j+1}, \ldots, \mathbf{a}_s$ $(1 < j < s)$ are linearly dependent.

3.2 Givens Transformations

The classical formulation of the Givens method (for example Gourlay and Watson, 1973) is to transform the two vectors – usually two rows of a matrix – by premultiplication by an elementary 2×2 matrix of the form

$$G = \begin{bmatrix} \cos\theta & \sin\theta \\ -\sin\theta & \cos\theta \end{bmatrix}.$$

Thus the problem is to find θ such that given

$$\mathbf{u}^{\mathrm{T}} = [0, \ldots, 0, u_s, \ldots, u_N], \qquad \mathbf{v}^{\mathrm{T}} = [0, \ldots, 0, v_s, \ldots, v_N],$$

$$G\begin{bmatrix} \mathbf{u}^{\mathrm{T}} \\ \mathbf{v}^{\mathrm{T}} \end{bmatrix} = \begin{bmatrix} 0, \ldots, 0, \bar{u}_s, & \ldots\ldots, \bar{u}_N \\ 0, \ldots, 0, 0, \bar{v}_{s+1}, & \ldots, \bar{v}_N \end{bmatrix}. \tag{3.7}$$

If $\mathbf{u}^{\mathrm{T}}$ and $\mathbf{v}^{\mathrm{T}}$ are the pth and qth rows respectively $(q > p)$ of an $N \times N$ matrix A, then the equivalent statement of (3.7) is

$$G^{+}A = \bar{A},$$

where A and $\bar{A}$ differ only in the pth and qth rows, and where

$$G^{+} = \begin{cases} g_{pp} = g_{qq} = \cos\theta \\ g_{pq} = -g_{qp} = \sin\theta \\ g_{ii} = 1, & i \neq p \neq q \\ g_{ij} = 0 & \text{otherwise.} \end{cases}$$

Thus it is easier to define the procedure in terms of the system (3.7), without altering the algebra involved. It follows from (3.7) that in order to obtain the additional zero in **v**, θ chosen is such that

$$-\sin\theta\, u_s + \cos\theta\, v_s = 0. \qquad (3.8a)$$

Since it is also true that

$$\cos\theta\, u_s + \sin\theta\, v_s = \bar{u}_s, \qquad (3.8b)$$

it follows that

$$\bar{u}_s = [u_s^2 + v_s^2]^{1/2} \qquad (3.8c)$$

and

$$\sin\theta = \frac{v_s}{\bar{u}_s}, \qquad \cos\theta = \frac{u_s}{\bar{u}_s}. \qquad (3.8d)$$

Thus for $i = s+1, \ldots, N$ it follows that

$$\bar{u}_i = \cos\theta\, u_i + \sin\theta\, v_i$$

$$= \frac{u_i u_s + v_i v_s}{\bar{u}} \qquad (3.8e)$$

and

$$\bar{v}_i = -\sin\theta\, u_i + \cos\theta\, v_i$$

$$= \frac{v_i u_s - u_i v_s}{\bar{u}}. \qquad (3.8f)$$

The natural way of performing the transformation therefore involves one square root and $6(N-s)+5$ arithmetic operations. It is the purpose of the so-called square-root-free Givens' transformations (due to Gentleman, 1973, see also Wilkinson, 1976) to considerably reduce this amount of arithmetic. The modification that avoids square roots does so in a manner not dissimilar to the practice of computing an LDL^T triangular factorization of a symmetric matrix rather than the LL^T Cholesky factorization, since the latter involves square roots. The only difference between the LDL^T factors and the LL^T factors is a scaling of the rows of L^T by the matrix $D^{1/2}$. Thus it might be expected that the exclusion of the square roots will be at the cost of some scaling, and this is indeed what happens.

Consider applying the Givens' transformation to the matrix KA, where $K = \text{diag}(k_i)$ corresponds to $D^{1/2}$ in the LDL^T factorization; again it is only necessary to consider matrices with two rows. The basic equations (3.7) are replaced by

$$G\begin{bmatrix} k_1 \mathbf{u}^T \\ k_2 \mathbf{v}^T \end{bmatrix} = \begin{bmatrix} 0, \ldots, 0, \bar{k}_1 \bar{u}_s, & \ldots, \bar{k}_1 \bar{u}_N \\ 0, \ldots, 0, 0, \bar{k}_2 \bar{v}_{s+1}, & \ldots, \bar{k}_2 \bar{v}_N \end{bmatrix},$$

$$= \begin{bmatrix} \bar{k}_1 \bar{\mathbf{u}}^T \\ k_2 \bar{\mathbf{v}}^T \end{bmatrix} \qquad (3.9)$$

nd the computation is carried out in such a way that only $k_1^2, k_2^2, \bar{k}_1^2$ and $\bar{k}_2^2$ need e stored, since explicit evaluation of the k_i would involve extracting square roots. follows from (3.9) that

$$\left.\begin{array}{l} -\sin\theta\, k_1 u_s + \cos\theta\, k_2 v_s = 0 \\ \text{nd} \\ \cos\theta\, k_1 u_s + \sin\theta\, k_2 v_s = \bar{k}_1 \bar{u}_s \end{array}\right\}. \tag{3.10}$$

hus for $i = s + 1, \ldots, N$,

$$\left.\begin{array}{l} -\sin\theta\, k_1 u_i + \cos\theta\, k_2 v_i = \bar{k}_2 \bar{v}_i \\ \text{nd} \\ \cos\theta\, k_1 u_i + \sin\theta\, k_2 v_i = \bar{k}_1 \bar{u}_i \end{array}\right\}. \tag{3.11}$$

f LDL^T factors are generated in place of LL^T factors, it is possible to use the xtra degrees of freedom to demand that the diagonal elements of L are unity. imilarly in the square-root-free Givens' transformations, it is possible to assume hat the extra degrees of freedom provided by K can be used to set

$$\left.\begin{array}{l} \bar{k}_1 = k_1 \cos\theta \\ \text{nd} \\ \bar{k}_2 = k_2 \cos\theta \end{array}\right\}. \tag{3.12a}$$

t then follows from (3.10) that

$$\sin\theta = \frac{\bar{k}_1 k_2}{k_1^2}\left(\frac{v_s}{u_s}\right) \tag{3.12b}$$

ence

$$\bar{k}_1^2\left(1 + \left(\frac{k_2}{k_1}\frac{v_s}{u_s}\right)^2\right) = k_1^2 \tag{3.13}$$

nd

$$\bar{k}_2^2\left(1 + \left(\frac{k_2}{k_1}\frac{v_s}{u_s}\right)^2\right) = k_2^2. \tag{3.14}$$

It is possible to rewrite formulae (3.10) and (3.11) in terms of $l_i \equiv k_i^2$ and $\equiv \bar{k}_i^2$ $(i = 1, 2)$, and to evaluate $\bar{v}_i, \bar{u}_i$ $(i = s + 1, \ldots, N)$ without the use of square oots. First it is clear that (3.13) and (3.14) can be written as

$$\bar{l}_1 = l_1/(1 + \alpha\beta) \tag{3.15a}$$

nd

$$\bar{l}_2 = l_2/(1 + \alpha\beta) \tag{3.15b}$$

here,

$$\alpha = \frac{v_s}{u_s} \tag{3.15c}$$

and

$$\beta = \frac{l_2}{l_1} \frac{v_s}{u_s}. \qquad (3.15$$

It then follows that

$$\bar{u}_i = u_i + \beta v_i, \qquad i = s, \ldots, N \qquad (3.15$$

and

$$\bar{v}_i = v_i - \alpha u_i, \qquad i = s + 1, \ldots, N. \qquad (3.15$$

These formulae (3.15a)–(3.15f) can then be evaluated by means of only $4(N - s) + 7$ arithmetic operations. Note that it is assumed that the transformatio matrix G is not required explicitly. This is not true for eigenproblems in which *eigenvectors* as well as eigenvalues are required, and such problems can lead to difficulties in implementing the square-root-free form of the Givens' transformatio *Bounds* on the round-off error propagated by the square-root-free Givens are superior for those of Householder's transformations; but it is not clear whether these bounds are a true reflection of the errors themselves.

Exercises

3.3. Verify that if

$$G\begin{bmatrix} k_1 \mathbf{u}^{\mathrm{T}} \\ k_2 \mathbf{v}^{\mathrm{T}} \end{bmatrix} = \begin{bmatrix} \bar{k}_1 \bar{\mathbf{u}}^{\mathrm{T}} \\ \bar{k}_2 \bar{\mathbf{v}}^{\mathrm{T}} \end{bmatrix}$$

where $\bar{k}_1$, $\bar{k}_2$, $\bar{\mathbf{u}}$ and $\bar{\mathbf{v}}$ are defined as above, then for any $\mathbf{p} = [p_1, \ldots, p_M]^{\mathrm{T}}$ and $\mathbf{q} = [q_1, \ldots, q_M]^{\mathrm{T}}$,

$$G\begin{bmatrix} \frac{1}{k_1}\mathbf{p}^{\mathrm{T}} \\ \\ \frac{1}{k_2}\mathbf{q}^{\mathrm{T}} \end{bmatrix} = \begin{bmatrix} \frac{1}{\bar{k}_1}\bar{\mathbf{p}}^{\mathrm{T}} \\ \\ \frac{1}{\bar{k}_2}\bar{\mathbf{q}}^{\mathrm{T}} \end{bmatrix}.$$

It follows that

$$\bar{p}_i = (p_i + \alpha q_i)/(1 + \alpha\beta)$$

and $\qquad i = 1, \ldots, M$

$$\bar{q}_i = (q_i - \beta p_i)/(1 + \alpha\beta).$$

This result is required if the methods of Section 7.3(a) are to be applied using square-root-free Givens' transformations.

3.4. Verify that the formulae (3.8) lead to $6(N - s) + 5$ operations, whereas formulae (3.15) lead to only $4(N - s) + 7$.

3.3 Overdetermined Systems: An Introduction

In Chapter 2 various methods are considered for solving a system of N linear quations in N unknowns; that is

$$A\mathbf{x} = \mathbf{b}$$

here A is an $N \times N$ matrix. An equally important problem is 'solving' M equations N unknowns ($M > N$). In general, given an $M \times N$ matrix A and a vector $= [b_1, \ldots, b_M]^{\mathrm{T}}$, there is no $\mathbf{x} = [x_1, \ldots, x_N]^{\mathrm{T}}$ such that $A\mathbf{x} = \mathbf{b}$; the 'solution' the vector $\mathbf{x}$ that minimizes the *residual vector*

$$\mathbf{r} \equiv \mathbf{b} - A\mathbf{x}.$$

The residual is minimized in terms of a norm such as:

(1) $\| \mathbf{r} \|_1 \equiv \sum_i | r_i |,$ (l_1 – norm)

(2) $\| \mathbf{r} \|_2 \equiv \{ \sum_i r_i^2 \}^{1/2} \equiv \{ \mathbf{r}^{\mathrm{T}} \mathbf{r} \}^{1/2}$ (l_2 – norm)

(3) $\| \mathbf{r} \|_\infty \equiv \max_i | r_i |.$ (l_∞ –norm)

ach one of these three leads to a solution with slightly different properties. A orough discussion of the relative merits of the different forms of approximation beyond the scope of this book and the reader is referred to Rice (1964). It has een suggested however that the type of approximation should be selected ccording to the nature of the data to be approximated:

l_∞ – for precise mathematical data,

l_2 – for 'noisy' statistical data,

l_1 – for data containing noise plus a few large errors.

In view of the almost universal use of l_2 when dealing with statistical data – or example regression analysis – hereafter that method alone is considered. The xclusion of the other two forms constitutes neither a judgement on their relative nerits nor an assessment of their applicability; it is simply that the methods would ot fit into the present text.

he method of least-squares (l_2 – approximation).

The least-squares solution of an overdetermined system $A\mathbf{x} = \mathbf{b}$ is the vector $\mathbf{x}^* = [x_1^*, \ldots, x_N^*]^{\mathrm{T}}$ such that

$$\| A\mathbf{x}^* - \mathbf{b} \|_2^2 = \min_{\mathbf{x}} \| A\mathbf{x} - \mathbf{b} \|_2^2.$$

t follows from elementary calculus that an equivalent statement of this condition s

$$\frac{\partial}{\partial x_i} \{ \| A\mathbf{x}^* - \mathbf{b} \|_2^2 \} = 0, \qquad i = 1, \ldots, N. \tag{3.16}$$

From the definition of the l_2-norm

$$\| A\mathbf{x} - \mathbf{b} \|_2^2 = (\mathbf{b}^T - \mathbf{x}^T A^T)(A\mathbf{x} - \mathbf{b})$$

and thus (3.16) can be written as

$$A^T A\mathbf{x}^* = A^T\mathbf{b} \tag{3.17}$$

these N equations in N unknowns are called the *normal equations.* The normal equations are usually not recommended for use in finding the l_2 solution of an overdetermined system, but their use has been combined with iterative refinement to give satisfactory solutions. The system (3.17) tends to be ill-conditioned; that is small changes in A and $\mathbf{b}$ lead to relatively large changes in the solution vector.

It is not possible to apply Gauss elimination to an overdetermined system, for this would lead to a system with a solution that depended on the ordering of the equations. The solution also changes if an individual row is scaled.

Example The least-squares solution of

$$\begin{aligned} 3x_1 + x_2 &= 2 \\ x_1 - 3x_2 &= 4 \\ x_1 - x_2 &= 0 \\ x_1 + x_2 &= 1 \end{aligned}$$

is $x_1 = \frac{11}{12}, x_2 = -\frac{9}{12}$, whereas after scaling the last row the least squares solution of

$$\begin{aligned} 3x_1 + x_2 &= 2 \\ x_1 - 3x_2 &= 4 \\ x_1 - x_2 &= 0 \\ 5x_1 + 5x_2 &= 5 \end{aligned}$$

is $x_1 = \frac{15}{12}, x_2 = -\frac{5}{12}$. This example can be viewed as an illustration of weighted least-squares approximation.

Exercise

3.5. For the above example, form the normal equations and hence verify the solutions.

Orthogonal transformations

It is however possible to transform the system using *orthogonal matrices* without affecting the solution. For

$$A^T A\mathbf{x} - A^T\mathbf{b} = (A^T Q)\, Q^T A\mathbf{x} - (A^T Q)\, Q^T\mathbf{b},$$

and hence (3.17) represents the normal equations for any set of equations of the form

$$Q^{\mathrm{T}}A\mathbf{x} = Q^{\mathrm{T}}\mathbf{b}, \tag{3.18}$$

where Q is an orthogonal matrix. It is a sequence of orthogonal transformations that forms the basis of one of the most efficient methods for least-squares problems (Golub, 1965). Owing to lack of space, competitive methods such as those based on modified Gram-Schmidt (Bjorck, 1967a) and on generalized inverses (Fletcher, 1970) will not be discussed.

It is shown in Section 3.1 that, for any $M \times N$ ($M \geqslant N$) matrix of rank N, there exists an $M \times N$ matrix P with orthonormal columns and an $N \times N$ upper triangular matrix R such that

$$A = PR. \tag{3.19}$$

Let Q be an orthogonal matrix with the first N columns as in the matrix P; then the remaining columns can be any vectors orthogonal to the columns of P, as they take no part in the numerical computation. Define the $M \times N$ matrix

$$U \equiv \begin{bmatrix} R \\ 0 \end{bmatrix} \tag{3.20}$$

that is, augment R by $M - N$ rows of zeros. With Q defined by (3.19) and U defined by (3.20), it follows that (3.18) becomes

$$U\mathbf{x} = Q^{\mathrm{T}}\mathbf{b}. \tag{3.21}$$

If we write the vector on the right-hand side as $\begin{bmatrix} \mathbf{p} \\ \mathbf{q} \end{bmatrix}$, where $\mathbf{p}$ represents the first N components and $\mathbf{q}$ the remaining $M - N$, then the least-squares solution of (3.21) is the solution of the $N \times N$ system

$$R\mathbf{x} = \mathbf{p}.$$

This reduced system can be solved by back substitution. The remaining terms on the right-hand side give the value of the minimum residual, that is;

$$\min_{\mathbf{x}} \| A\mathbf{x} - \mathbf{b} \|_2 = \| \mathbf{q} \|_2 .$$

A more detailed discussion of least-squares problems can be found in Lawson and Hanson (1974).

3.4 Overdetermined Systems: A Numerical Implementation

Any method that reduces $A\mathbf{x} = \mathbf{b}$ to $R\mathbf{x} = \mathbf{p}$ by orthogonal transformations is in theory valid. Efficient sequences of transformations are provided by Householder's transformations (Powell and Reid, 1969) or Givens' transformations. In this section we describe the use of the latter as applied to the overdetermined system given in the previous section. We use square-root-free Givens' transformations to reduce the matrix of coefficients to upper-triangular form. Since the same transformations have to be applied to the vector on the right-hand side, the calculation is arranged so that the transformations are applied to the augmented matrix $[A \mid \mathbf{b}]$. Thus in this example, the matrix

$$[A \mid \mathbf{b}] = \left[\begin{array}{cc|c} 3 & 1 & 2 \\ 1 & -3 & 4 \\ 1 & -1 & 0 \\ 5 & 5 & 5 \end{array}\right] \tag{3.22}$$

is transformed into a matrix of the form

$$\left[\begin{array}{cc|c} * & * & * \\ & * & * \\ & & * \\ & & * \end{array}\right], \tag{3.23}$$

where * denotes a non-zero component. We write

$$[A \mid \mathbf{b}] = D^{1/2}G,$$

where D is the identity matrix. Givens' transformations are now applied to reduce the matrix G to the desired form (3.23).

In general the components below the diagonal are eliminated systematically – removing successive columns by transforming the appropriate rows. Thus the pth row of the first column ($p = 2, \ldots, M$) is removed by transforming rows 1 and p; then the second column is reduced by transforming 2nd and qth rows ($q = 3, \ldots, M$); and so on.

Numerical example (Solution of a least-squares problem using square-root-free Givens' transformations).

The initial system can be written as

$$\begin{bmatrix} 1 & & & \\ & 1 & & \\ & & 1 & \\ & & & 1 \end{bmatrix}^{1/2} \begin{bmatrix} 3 & 1 & 2 \\ 1 & -3 & 4 \\ 1 & -1 & 0 \\ 5 & 5 & 5 \end{bmatrix}.$$

Applying a transformation to the first two rows leads to the system

$$\begin{bmatrix} \frac{9}{10} & & & \\ & \frac{9}{10} & & \\ & & 1 & \\ & & & 1 \end{bmatrix}^{1/2} \begin{bmatrix} \frac{10}{3} & 0 & \frac{10}{3} \\ 0 & -\frac{10}{3} & \frac{10}{3} \\ 1 & -1 & 0 \\ 5 & 5 & 5 \end{bmatrix} \quad \begin{array}{l} (\bar{g}_{1i} = g_{1i} + \frac{1}{3}g_{2i}) \\ (\bar{g}_{2i} = g_{2i} - \frac{1}{3}g_{1i}) \\ (\bar{g}_{3i} = g_{3i}) \\ (\bar{g}_{4i} = g_{4i}). \end{array}$$

A transformation of the first and third rows leads to:

$$\begin{bmatrix} \frac{9}{11} & & & \\ & \frac{9}{10} & & \\ & & \frac{10}{11} & \\ & & & 1 \end{bmatrix}^{1/2} \begin{bmatrix} \frac{11}{3} & -\frac{1}{3} & \frac{10}{3} \\ 0 & -\frac{10}{3} & \frac{10}{3} \\ 0 & -1 & -1 \\ 5 & 5 & 5 \end{bmatrix} \quad \begin{matrix} (\bar{\bar{g}}_{1i} = \bar{g}_{1i} + \frac{1}{3}\bar{g}_{3i}) \\ (\bar{\bar{g}}_{2i} = \bar{g}_{2i}) \\ (\bar{\bar{g}}_{3i} = \bar{g}_{3i} - \frac{3}{10}\bar{g}_{1i}) \\ (\bar{\bar{g}}_{4i} = \bar{g}_{4i}). \end{matrix}$$

A transformation of the first and fourth rows leads to

$$\begin{bmatrix} \frac{1}{4} & & & \\ & \frac{9}{10} & & \\ & & \frac{10}{11} & \\ & & & \frac{11}{36} \end{bmatrix}^{1/2} \begin{bmatrix} 12 & 8 & \frac{35}{3} \\ 0 & -\frac{10}{3} & \frac{10}{3} \\ 0 & -1 & -1 \\ 0 & \frac{60}{11} & \frac{5}{11} \end{bmatrix} \quad \begin{matrix} (\bar{\bar{\bar{g}}}_{1i} = \bar{\bar{g}}_{1i} + \frac{5}{3}\bar{\bar{g}}_{4i}) \\ (\bar{\bar{\bar{g}}}_{2i} = \bar{\bar{g}}_{2i}) \\ (\bar{\bar{\bar{g}}}_{3i} = \bar{\bar{g}}_{3i}) \\ (\bar{\bar{\bar{g}}}_{4i} = \bar{\bar{g}}_{4i} - \frac{15}{11}\bar{\bar{g}}_{1i}). \end{matrix}$$

Then it is necessary to eliminate the second column.

A transformation of the second and third rows leads to

$$\begin{bmatrix} \frac{1}{4} & & & \\ & \frac{33}{40} & & \\ & & \frac{10}{12} & \\ & & & \frac{11}{36} \end{bmatrix}^{1/2} \begin{bmatrix} 12 & 8 & \frac{35}{3} \\ 0 & -\frac{40}{11} & \frac{100}{33} \\ 0 & 0 & -2 \\ 0 & \frac{60}{11} & \frac{5}{11} \end{bmatrix}.$$

Finally a transformation of the second and fourth rows leads to

$$\begin{bmatrix} \frac{1}{4} & & & \\ & \frac{9}{20} & & \\ & & \frac{10}{12} & \\ & & & \frac{1}{6} \end{bmatrix}^{1/2} \begin{bmatrix} 12 & 8 & \frac{35}{3} \\ 0 & -\frac{20}{3} & \frac{25}{9} \\ 0 & 0 & -2 \\ 0 & 0 & 5 \end{bmatrix}.$$

The desired least-squares solution is therefore the solution of the triangular system

$$\begin{bmatrix} 12 & 8 \\ & -\frac{20}{3} \end{bmatrix} \begin{bmatrix} x_1 \\ x_2 \end{bmatrix} = \begin{bmatrix} \frac{35}{3} \\ \frac{25}{9} \end{bmatrix},$$

which is solved (by substitution) to give $x_2 = -\frac{5}{12}$, $x_1 = \frac{15}{12}$. The non-zero terms in the remaining rows give the minimum *sum of squares of the residuals* i.e.,

$$\min_{\mathbf{x}} \| A\mathbf{x} - \mathbf{b} \|_2^2 = \tfrac{10}{12} \cdot 2^2 + \tfrac{1}{6} \cdot 5^1 = \tfrac{45}{6}.$$

In practical problems, the efficiency of least-squares methods can be enhanced by the use of iterative refinement (Björck, 1976b). Another feature of this form of least-squares solution using Givens' rotations is that additional equations can be added to the system without the necessity of repeating all the earlier transformations (see Exercise 3.7).

Exercises

3.6. Perform the transformations on the following system of equations:

$$\begin{aligned} 3x_1 + x_2 &= 2 \\ x_1 - 3x_2 &= 4 \\ x_1 - x_2 &= 0 \\ x_1 + x_1 &= 1. \end{aligned}$$

Verify that the least-squares solution is $x_1 = \frac{11}{12}$, $x_2 = -\frac{9}{12}$. What is the l_2-norm of the residual?

3.7. Add the additional row

$$4x_1 + 4x_2 = 4$$

to the system in Exercise 3.6 and verify that only two *additional* transformations are required to convert the solution of Exercise 3.6 into the solution of the worked example.

4

Sparse-matrix Techniques

4.1 Introduction

A *sparse* matrix is one in which zero components predominate; if such a matrix is also *large*, then it is reasonable to attempt to make use of the sparsity. Thus it may be possible to eliminate arithmetic operations performed on zeros and so speed up the solution of the problem.

A simple example of this technique is illustrated by the use of Gauss elimination to solve a tridiagonal system of equations which can be written as $A\mathbf{x} = \mathbf{b}$ where

$$A \equiv \begin{bmatrix} \alpha_1 & \gamma_1 & & & \\ \beta_1 & \alpha_2 & \gamma_2 & & \\ & \cdot & \cdot & \cdot & \\ & & \cdot & \cdot & \gamma_{N-1} \\ & & & \beta_{N-1} & \alpha_N \end{bmatrix}.$$

If it is not necessary to introduce any row interchanges – say for example if $|\alpha_i| > |\gamma_{i-1}| + |\beta_i|$ for all i (see Exercise 2.7) – then Gauss elimination can be written as

$$\alpha_1^* = \alpha_1, \qquad b_1^* = b_1,$$

$$\left.\begin{aligned} m_k &= \frac{\beta_{k-1}}{\alpha_{k-1}^*} \\ \alpha_k^* &= \alpha_k - m_k\gamma_{k-1} \\ b_k^* &= b_k - m_k b_{k-1}^* \end{aligned}\right\} \quad k = 2, \ldots, N.$$

Back substitution becomes

$$x_N = \frac{b_N^*}{\alpha_N^*}$$

$$x_k = \frac{b_k^* - \gamma_k x_{k+1}}{\alpha_k^*}, \qquad k = N-1, \ldots, 1.$$

This is an algorithm that not only avoids unnecessary arithmetic operations on zeros but also reduces the amount of storage required from $O(N^2)$ to approximately $5N$ or less, as the matrix A is replaced by vectors $\boldsymbol{\alpha}$, $\boldsymbol{\beta}$ and $\boldsymbol{\gamma}$.

This type of reduction can be efficiently implemented when the matrix concerned has a regular structure. In particular, the algorithm for elimination in tridiagonal matrices is used to solve large algebraic problems that arise as finite-difference approximations to partial differential equations in several dimensions using block iterative methods (see Section 5.5). Direct methods for structured matrices are considered in Section 4.3.

In many problems however, there is not this degree of regularity and so more general techniques applicable to matrices with arbitrary structure are discussed in Section 4.2. The object of such methods is, as above, twofold; to speed up the calculation by avoiding arithmetic operations on zeros and to use minimum space by not storing zeros. By so doing, it is possible to solve bigger problems, and to solve them faster.

If only the non-zero components and their positions in the matrix are to be stored, a problem arises as to the most effective method of storing such information. As any detailed comparison of such storage schemes enters the realms of computer science, it is beyond the scope of the present text; two possible alternatives are given without any serious assessment of their relative merits. Alternative schemes that take into account the structure of a sparse matrix are outlined in Section 4.3.

(1) *A linked list* The list is made up of units, each containing

(a) the position of a non-zero component – possibly as a pair of subscripts (i, j);
(b) the value of the component;
(c) a pointer to the next unit.

Such a list is illustrated schematically in Figure 2(a).

(2) *A binary map* The structure of the matrix is stored in a condensed form as a pattern of *bi*nary digi*ts* (*bits*) known as an *adjacency matrix*. Thus a whole row may be stored in a location that would normally hold a single number. The values of the non-zero components would then be stored row by row in a single vector. For example the matrix

$$A = \begin{bmatrix} 6 & 2 & 0 & 0 & 4 \\ 1 & -5 & 0 & 1 & 0 \\ 0 & 1 & 4 & 0 & 0 \\ 3 & 0 & 1 & 0 & 0 \\ 0 & 4 & 0 & 0 & 7 \end{bmatrix} \tag{4.1}$$

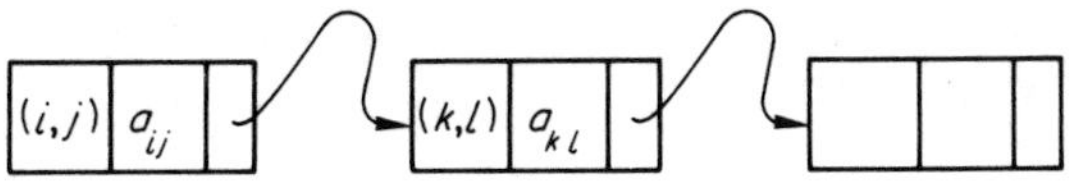

(a)

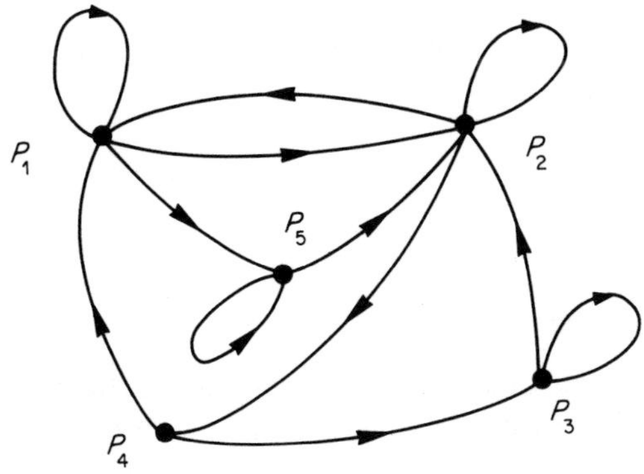

(b)

Figure 2 (a) A linked list; (b) the directed graph $\mathscr{C}(A)$

would be stored as

$$B(A) = \begin{bmatrix} 1 & 1 & 0 & 0 & 1 \\ 1 & 1 & 0 & 1 & 0 \\ 0 & 1 & 1 & 0 & 0 \\ 1 & 0 & 1 & 0 & 0 \\ 0 & 1 & 0 & 0 & 1 \end{bmatrix},$$

which requires only 25 bits, together with the numerical values of the non-zeros:

$$\mathbf{V}(A) = [6, 2, 4, 1, -5, 1, 1, 4, 3, 1, 4, 7]^{\mathrm{T}}.$$

Sometimes the binary (or *boolean*) adjacency matrix is interpreted as a *directed graph* $\mathscr{C}(A)$, where $b_{ij} = 1$ indicates a path from P_i to P_j; the graph, corresponding to (4.1), is shown in Figure 2(b).

The major difficulty in implementing either of these condensed forms of storage is that during the elimination procedure, components that are initially zero do not necessarily remain zero; this is known as *fill-in*.

For example, with the matrix given in (4.1), partial pivoting does not alter the order and it follows that with $A^{(1)} = A$,

$$B(A^{(2)}) = \begin{bmatrix} 1 & 1 & 0 & 0 & 1 \\ 0 & 1 & 0 & 1 & (1) \\ 0 & 1 & 1 & 0 & 0 \\ 0 & (1) & 1 & 0 & (1) \\ 0 & 1 & 0 & 0 & 1 \end{bmatrix},$$

where the 'new' non-zero components are indicated by (1). Similarly

$$B(A^{(3)}) = \begin{bmatrix} 1 & 1 & 0 & 0 & 1 \\ 0 & 1 & 0 & 1 & 1 \\ 0 & 0 & 1 & (1) & (1) \\ 0 & 0 & 1 & (1) & 1 \\ 0 & 0 & 0 & (1) & 1 \end{bmatrix},$$

after which clearly no further fill-in is possible. While the methods outlined in the subsequent sections are illustrated by simple examples such as the one above, the true worth of the methods does not become apparent unless much larger problems are tackled.

Special techniques for sparse matrices have been the objects of detailed study for some time and the subject of discussion at several conferences (for example, Reid, 1971a; Rose and Willoughby, 1972; Bunch and Rose, 1976). It is possible therefore to give only a very brief outline of some of the more fruitful avenues of study, and to present a selection of a few of the more significant results.

It should be noted that the use of boolean matrices to illustrate the examples here does not indicate a preference for that method of representation in a computer. As a general rule a sparse system warrants special treatment if there are not more than 10% of the components non-zero. If several systems with the same structure are to be solved, this proportion could be as high as 50%.

Exercise

4.1. Verify that the Crout factorization leads to the same fill-in as Gauss elimination if the same pivoting strategy is used.

4.2. Reduction of Fill-in

In this section four methods are considered which have all been suggested as suitable for sparse matrix problems. Together they illustrate the trade-off between obtaining an efficient algorithm for eliminating a sparse matrix and the effort required in devising the pivoting strategy – known as *preprocessing*.

(a) Minimizing the local fill-in

In this method the pivot at the kth stage is chosen such that the fill-in of the matrix $B^{(k)}$ is minimized.

For any boolean matrix B it is possible to define a *matrix complement* $\bar{B}$ such that if

$$B = \{b_{ij}\} \qquad \text{and} \qquad \bar{B} = \{\bar{b}_{ij}\},$$

then

$$\bar{b}_{ij} \equiv 1 - b_{ij}.$$

Theorem (Tewarson, 1970) The fill-in that results from the use of $a_{ij}^{(k)}$ $(\neq 0)$ $i, j \geqslant k$ *as pivot at the kth stage of the elimination of A is given by the corresponding component* $g_{ij}^{(k)}$ *of the* local-cost matrix

$$G^{(k)} = B^{(k)} \bar{B}^{(k)\mathrm{T}} B^{(k)},$$

where

$$B^{(k)} = B(C^{(k)}). \quad \Delta$$

Proof When $a_{ij}^{(k)}$ is used as pivot, the Gauss elimination formula can be written as

$$a_{pq}^{(k+1)} = a_{pq}^{(k)} - \frac{a_{pj}^{(k)} a_{iq}^{(k)}}{a_{ij}^{(k)}} \qquad \{p \neq i, q \neq j, \quad p, q \geqslant k\}.$$

Thus fill-in of the (p, q) component occurs if and only if

$$a_{pq}^{(k)} = 0, a_{pj}^{(k)} \neq 0 \neq a_{iq}^{(k)};$$

that is, if

$$b_{pj}^{(k)} = b_{iq}^{(k)} = 1 - b_{pq}^{(k)} = 1.$$

Thus as

$$1 - b_{pq}^{(k)} \equiv \bar{b}_{pq}^{(k)},$$

it follows that the total cost of $a_{ij}^{(k)}$ – the fill-in resulting from $a_{ij}^{(k)}$ as pivot – is

$$g_{ij}^{(k)} = \sum_{\substack{p=k \\ p \neq i}}^{N} \sum_{\substack{q=k \\ q \neq j}}^{N} b_{iq}^{(k)} \bar{b}_{pq}^{(k)} b_{pj}^{(k)}. \tag{4.2}$$

Since

$$b_{iq}^{(k)} \bar{b}_{iq}^{(k)} b_{ij}^{(k)} = 0, \qquad \text{for all } q$$

and

$$b_{ij}^{(k)} \bar{b}_{pj}^{(k)} b_{pj}^{(k)} = 0, \qquad \text{for all } p,$$

the restrictions $p \neq i, q \neq j$ can be deleted. Thus the right-hand side of (4.2) becomes the (i, j) component of $B^{(k)}\bar{B}^{(k)\mathrm{T}}B^{(k)}$, which is the desired result. Δ

However, it has been shown by counter example (Duff and Reid, 1974) that the strategy of selecting the pivot with minimum local cost does not necessarily lead to the minimum total fill-in.

Numerical example (The use of the local-cost matrix to determine a pivoting strategy). Given the matrix (4.1), $B^{(1)} = B(A)$; hence

$$\bar{B}^{(1)} = \begin{bmatrix} 0 & 0 & 1 & 1 & 0 \\ 0 & 0 & 1 & 0 & 1 \\ 1 & 0 & 0 & 1 & 1 \\ 0 & 1 & 0 & 1 & 1 \\ 1 & 0 & 1 & 1 & 0 \end{bmatrix}$$

and

$$G^{(1)} = \begin{bmatrix} 3 & 4 & 4 & 1 & 1 \\ 3 & 5 & 4 & 0 & 3 \\ 3 & 3 & 1 & 1 & 2 \\ 2 & 5 & 1 & 1 & 3 \\ 3 & 2 & 3 & 1 & 0 \end{bmatrix}.$$

There are therefore two possible pivots that lead to no fill-in at the first stage. If we select a_{24} as the first pivot it follows that

$$G^{(2)} = \begin{bmatrix} 2 & 3 & 4 & & 1 \\ & & & & \\ 2 & 2 & 1 & & 2 \\ 1 & 3 & 1 & & 3 \\ 2 & 1 & 3 & & 0 \end{bmatrix} \quad (4 \times 4 \text{ matrix}).$$

Selecting a_{55} as pivot at the second stage again leads to no fill-in; thus

$$B^{(3)} = \begin{bmatrix} 1 & 1 & 0 \\ 0 & 1 & 1 \\ 1 & 0 & 1 \end{bmatrix}$$

and

$$G^{(3)} = \begin{bmatrix} 1 & 1 & 2 \\ 2 & 1 & 1 \\ 1 & 2 & 1 \end{bmatrix}.$$

Any of the six possible pivots leads to the fill-in of one component. Thereafter no further fill-in is possible.

Exercise

4.2. Verify that the local cost of a_{ij} cannot increase if there is no fill-in of the ith row and jth column.

(b) Markowitz criterion

An alternative strategy which avoids the large preprocessing effort of the previous method was first suggested by Markowitz (1957). If θ_i is the number of non-zero components in the ith row of $B^{(k)}$ and ϕ_j is the number of non-zero components in the jth column, then the cost of $a_{ij}^{(k)}$ – assumed non-zero – as the kth pivot is not greater than $(\theta_i - 1)(\phi_j - 1)$. Thus a reasonable strategy is to select $a_{pq}^{(k)}$ $(\neq 0)$ as pivot, such that

$$(\theta_p - 1)(\phi_q - 1) = \min_{a_{ij}^{(k)} \neq 0} \{(\theta_i - 1)(\phi_j - 1)\}.$$

Numerical example (The use of the Markowitz criterion to determine a pivoting strategy). For the matrix (4.1), the values of $(\theta_i - 1)(\phi_j - 1)$ are initially given by the matrix

$$\Theta^{(1)} = \begin{bmatrix} 4 & 6 & 2 & 0 & 2 \\ 4 & 6 & 2 & 0 & 2 \\ 2 & 3 & 1 & 0 & 1 \\ 2 & 3 & 1 & 0 & 1 \\ 2 & 3 & 1 & 0 & 1 \end{bmatrix}.$$

Thus this strategy would choose a_{24} as pivot, as it leads to no fill-in. At the second stage–deleting row 2 and column 4 leads to

$$\Theta^{(2)} = \begin{bmatrix} 2 & 4 & 2 & & 2 \\ & & & & \\ 1 & 2 & [1] & & 1 \\ [1] & 2 & [1] & & 1 \\ 1 & 2 & 1 & & [1] \end{bmatrix},$$

where square-bracketed numbers denote $a_{ij}^{(2)} \neq 0$, and so there are four possible pivots to choose from. If $a_{41}^{(2)}$ is taken as the pivot, then this leads to

$$B^{(3)} = \begin{bmatrix} 1 & (1) & 1 \\ 1 & 1 & 0 \\ 1 & 0 & 1 \end{bmatrix},$$

where parentheses indicate fill-in, and

$$\Theta^{(3)} = \begin{bmatrix} 4 & 2 & 2 \\ 2 & [1] & 1 \\ 2 & 1 & [1] \end{bmatrix}.$$

Clearly there can be no further fill-in.

Examples have been constructed to show that it is not possible to state in advance which of the two strategies will be the best (Duff and Reid, 1974). Thus the Markowitz criterion is often preferred on the basis of simplicity and speed.

(c) Other methods

It is possible to define pivoting strategies that involve less preprocessing, but almost inevitably there is a reduction in effectiveness. A simple approach is to specify an ordering of the columns in advance and only select row interchanges thereafter (Curtis and Reid, 1972).

The ordering of the columns is usually based on values of ϕ_j for the original matrix; an alternative criterion which leads to marginally different results is the *density measure*:

$$\delta_j = \sum_{\substack{i=1 \\ a_{ij} \neq 0}}^{N} \theta_i, \qquad j = 1, \ldots, N.$$

If the columns are interchanged at each stage on the basis of updated values of ϕ_j or δ_j, then the method is comparable in complexity to the Markowitz method and does not provide any better results (Tosovic, 1973). The pivot row is then selected as the row of $B^{(k)}$ with minimum θ_i, subject to the pivot being non-zero. An even simpler selection procedure is to choose the pivot row on the basis of the θ_i values for the original matrix, then the pivots are all specified in advance. Duff and Reid (1974) have shown that this leads to greater fill-in, but it can be significantly easier to implement. In *all* the methods described in Section 4.3 *all* the pivots are specified in advance, but the matrices then are *very* large structured, symmetric and invariably positive definite.

All the interchange strategies can lead to problems with round-off errors unless they are combined with some criterion that specifies a lower bound on the size of the pivot. If such a mixed strategy is used, it is no longer possible to define the pivoting strategy in advance using only the structure of the matrix. Thus it is no longer possible to determine in advance the amount of storage required. There are two possible tests for acceptability of a pivot: (1) absolute size, and (2) relative size. A criterion based on the former might be that the absolute value of the pivot had to be greater than some small threshold value. A criterion based on relative size would be similar to the pivot selection described in Section 2.2; namely that $a_{ij}^{(k)}$ was an

:ceptable pivot if and only if

$$|a_{ij}^{(k)}| \geqslant \alpha \max_{k \leqslant p \leqslant N} |a_{pj}^{(k)}|, \tag{4.3}$$

here $\alpha = \frac{1}{4}$ and $\alpha = \frac{1}{100}$ have both been suggested as sensible choices.

It is often useful to investigate the possibility of reordering a sparse matrix into ɔck triangular form; that is, considering whether the matrix is reducible. The ork involved in such a test is often relatively small and the savings in computer me solving block triangular systems is considerable (Duff, 1976).

An alternative introduction to general sparse matrix techniques can be found in rameller *et al.* (1976).

Exercise

3(a) Verify that if the criterion (4.3) is used then

$$\max_{i,j \geqslant k} |a_{ij}^{(k+1)}| \leqslant (1 + \alpha^{-1}) \max_{i,j \geqslant k} |a_{ij}^{(k)}|.$$

(b) Show that

$$\max_{i,k} |a_{ij}^{(k)}| \leqslant (1 + \alpha)^{p_j - 1} \max_i |a_{ij}|,$$

here p_j is the final number of non-zero components in the jth column.

4.3 Methods for Structured Matrices

This section is concerned with methods for matrices arising from grid ɔproximations such as finite elements or finite differences. Iterative methods ve proven very useful for finite-difference methods in particular (see Mitchell,)69), they are discussed in this context in Section 5.5. Finite-element methods ee Mitchell and Wait, 1977), on the other hand, have tended to be solved by direct ethods and a few of the alternative approaches are described in this section. It is ɔt necessary to explain in detail how the algebraic system arises; it is only ecessary to observe that the approximate solutions of partial differential equations ch as (1.9) satisfy large sparse systems of equations. The graph of the matrix is irectly related to the grid of *elements* into which the domain of the problem is ıbdivided. As the matrices are usually symmetric and positive definite it is possible) store all the necessary information in an *undirected* graph. This graph is then sed to determine *a priori* the sequence of pivots. Since the Cholesky factorization ee Section 2.4) is used it is necessary to preserve the symmetry and so the 'mmetric row and column interchanges can be represented as a simple reordering f the nodes of a graph.

If the domain of a differential equation is partitioned into many small elements ay triangles or quadrilaterals), then in a simple approximation the unknowns are e approximate values at the nodes (the corners of the elements). It then follows

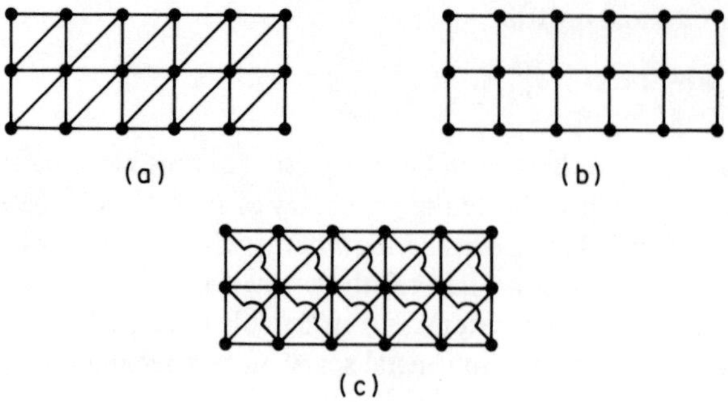

Figure 3

that in a finite-element approximation, a_{ij} can be non-zero only if nodes p_i and p_j are in the same element. Thus, for example, the graph corresponding to a simple triangular grid in two dimensions (Figure 3(a)) is identical to the network of grid lines, whereas the graph corresponding to a rectangular grid (Figure 3(b)) contains overlapping diagonals (Figure 3(c)). This latter will, in general, be a non-planar graph.

The object of the methods of this section is to use the structure of the matrix t reduce significantly the amount of preprocessing involved and still provide a good method.

(a) Reverse Cuthill–McKee

It is assumed for the purposes of this method that it is possible to provide a Cholesky factorization algorithm for the *envelope* of a matrix and in so doing the store required is related to the size of the *profile*.

For any matrix $A = \{a_{ij}\}$, let

$$f_i \equiv \min_j \{j : a_{ij} \neq 0\};$$

then it is possible to define

(1) $\text{Bandwidth} \equiv \max_i \{i - f_i\}$

(2) $\text{Profile} \equiv \sum_i (i - f_i)$

(3) $\text{Band}(A) \equiv \{(i,j) : 0 < i - j \leqslant \text{bandwidth}\}$ (4.

and

(4) $\text{Envelope}(A) \equiv \{(i,j) : f_i \leqslant j < i\}$.

Clearly if the matrix has only a few non-zeros in each row and if they are clustered around the diagonal, then a solution procedure that needs to store only

diagonal terms together with Band (A) is more efficient than one that stores all
l and replacing Band (A) by Envelope (A) is even more efficient.

As the methods introduced in this section can all be viewed most easily in terms graphs it is necessary to introduce a few more graph-theoretic concepts. Two les of a graph are said to be adjacent if there is *an edge* joining them; then for any le p,

$$\text{Adj}\,(p) \equiv \{q : p \text{ and } q \text{ are adjacent}\}$$

l for any set X of nodes,

$$\text{Adj}\,(X) = \bigcup_{p \in X} \text{Adj}\,(p) \backslash X. \tag{4.5}$$

e degree of p, that is $|\,\text{Adj}\,(p)\,|$ is the size of this set of adjacent nodes. For any le p, the *rooted level structure* $\mathcal{L}(p)$ is a partitioning of the nodes of the graph o sets $L_i \equiv L_i(p)$ ($i = 0, \ldots, l(p)$), where

$$L_0 \equiv \{p\},$$
$$L_1 = \text{Adj}\,(p)$$

d

$$L_{i+1} = \text{Adj}\,(L_i) \backslash L_{i-1} \qquad i = 1, 2, \ldots, l(p) - 1.$$

orithm for Reverse Cuthill–McKee

(1) Determine a *good starting node* p_1;
(2) for $i = 1, 2, \ldots, N$, find the unnumbered neighbours of node p_i and number m in increasing order of degree;
(3) invert the order.

te that (1) plus (2) gives the Cuthill–McKee ordering (Cuthill and McKee, 1969) l that step (2) is equivalent to constructing the rooted level structure $\mathcal{L}(p_1)$ and mbering the nodes level by level in a particular way. A good starting node would one that leads to a rooted level structure with as many levels as possible. An orithm for finding such a node has been provided by Gibbs *et al.* (1976).

gorithm for finding a starting node

(1) Find a node p, of minimal degree (easy);
(2) generate

$$\mathcal{L}(p) \equiv \{L_0(p), \ldots, L_{l(p)}(p)\};$$

(3) sort $L_{l(p)}(p)$ in order of increasing degree;
(4) then for each $q \in L_{l(p)}(p)$, generate $\mathcal{L}(q)$, and if $l(q) > l(p)$ replace p by q d return to (2);
(5) the final node p is a so-called *pseudo-peripheral node* and will be a good rting node.

The reason for inverting the order in step (3) of the reverse Cuthill–McKee algorithm is that while both the forward and reverse orderings provide the same bandwidth, the reverse ordering frequently provides a smaller envelope. It can never increase the envelope. The proof of this observation is due to Liu and Sherman (1976). Corresponding to definitions (4.4), which used the number of non-zeros on a row, define

$$h_j \equiv \max_i \{j : a_{ij} \neq 0\}$$

$$\text{T-profile} \equiv \sum_j (h_j - j)$$

and

$$\text{T-envelope}(A) \equiv \{(i,j) : j < i \leqslant h_j\}.$$

In addition, denote by A_{CM} the matrix A reordered according to the Cuthill–McKee algorithm and by A_{RCM} the matrix A subject to the reverse ordering. The for any A

$$\text{Envelope}(A_{\mathrm{CM}}) \equiv \text{T-envelope}(A_{\mathrm{RCM}}). \quad (4$$

Theorem (1) *$i \leqslant j$ implies $f_i \leqslant f_j$ for any A_{CM};*
(2) *T-envelope(A_{CM}) $\subset$ Envelope(A_{CM});*
and finally the main result

(3) *Envelope(A_{RCM}) $\subset$ Envelope(A_{CM})*

and

Profile(A_{RCM}) $\leqslant$ Profile(A_{CM}). Δ

Proof (1) $f_i \leqslant f_j$

if and only if

$$\min_k \{k : p_k \in \text{Adj}(p_i)\} \leqslant \min_l \{l : p_l \in \text{Adj}(p_j)\}.$$

This follows directly since $i \leqslant j$ implies that p_i precedes p_j, and the neighbours of p_i must precede the neighbours of p_j.

(2) Assume the contrary, that is there exists (i,j) such that

$$(i,j) \in \text{T-envelope}(A_{\mathrm{CM}}) \quad (4.$$

and

$$(i,j) \notin \text{Envelope}(A_{\mathrm{CM}}). \quad (4.$$

It follows from (4.7) that there exists $k \geqslant i$ such that $a_{kj} \neq 0$, and from (4.8) that for all $l \leqslant j, a_{il} = 0$. It then follows that there exists $k > i$ such that $f_k < f_i$ (see Figure 4).

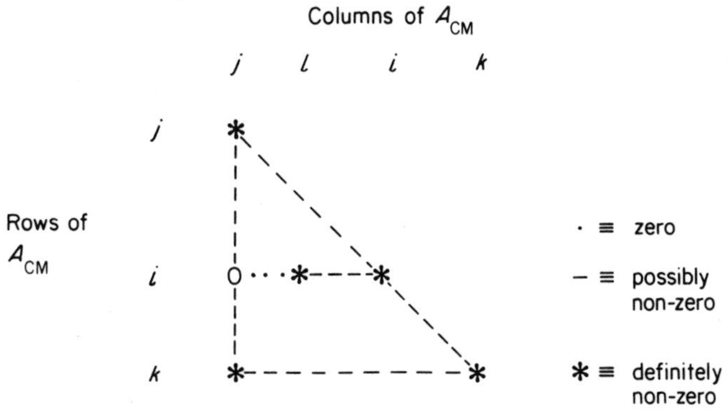

Figure 4

ut it follows from (1) that this is a contradiction and this part of the proof is omplete.

(3) This follows directly from (2) and (4.6). Δ

ote that

$$\text{Envelope}(A_{RCM}) \subset \text{Envelope}(A_{CM})$$

hould be interpreted as $(i, j) \in \text{Envelope}(A_{RCM})$ implies $(N+1-j, N+1-i) \in$ nvelope(A_{CM}).

If the entire envelope of a matrix is stored it is not necessary to use the linked sts or bit maps of the previous section. It is possible to provide all the data and orage necessary for the Cholesky factorization in two simple arrays. The first rray contains the matrix components (of the lower triangle) row by row and the econd array contains pointers to the entries of the first array that correspond to he diagonal terms.

xample For the particular grid illustrated, Figure 5(a) gives the Cuthill–McKee rdering and Figure 5(b) gives the reverse Cuthill–McKee ordering. The rooted vel structure corresponding to node p_1 in Figure 5(a) is

$$\mathcal{L}(p_1) = \{L_0, L_1, L_2, L_3, L_4, L_5, L_6\},$$

here

$$L_0 \equiv \{p_1\}, \quad L_1 \equiv \{p_2, p_3\}, \quad L_2 \equiv \{p_4, p_5\}, \quad L_3 \equiv \{p_6, p_7, p_8, p_9\},$$

$$L_4 \equiv \{p_{10}, p_{11}, p_{12}\}, \quad L_5 \equiv \{p_{13}, p_{14}, p_{15}\}, \quad L_6 \equiv \{p_{16}\}.$$

ince $l(p_1) = 6 = l(p_{16})$, it follows that p_1 is a pseudo-peripheral node and can be sed as a starting node for a Cuthill–McKee ordering.

For the two orderings shown,

$$\text{Bandwidth}(A_{CM}) = \text{Bandwidth}(A_{RCM}) = 5,$$

$$\text{Profile}(A_{CM}) = 41$$

and

$$\text{Profile}(A_{\text{RCM}}) = 34.$$

The matrix A_{RCM} is stored as

$$\mathbf{V} = [a_{11}, a_{22}, a_{21}, a_{33}, a_{32}, a_{31}, a_{44}, a_{55}, a_{54}, a_{53}, a_{52}, a_{66}, a_{65}, a_{64}, a_{63}, a_{77}, \ldots]^{\text{T}}$$

together with the pointers to the diagonal terms

$$\mathbf{D} = [1, 2, 4, 7, 8, 12, 16, \ldots]^{\text{T}}.$$

The storage required for the first array is profile $+ N$ (= 50); for the second array it is N (= 16).

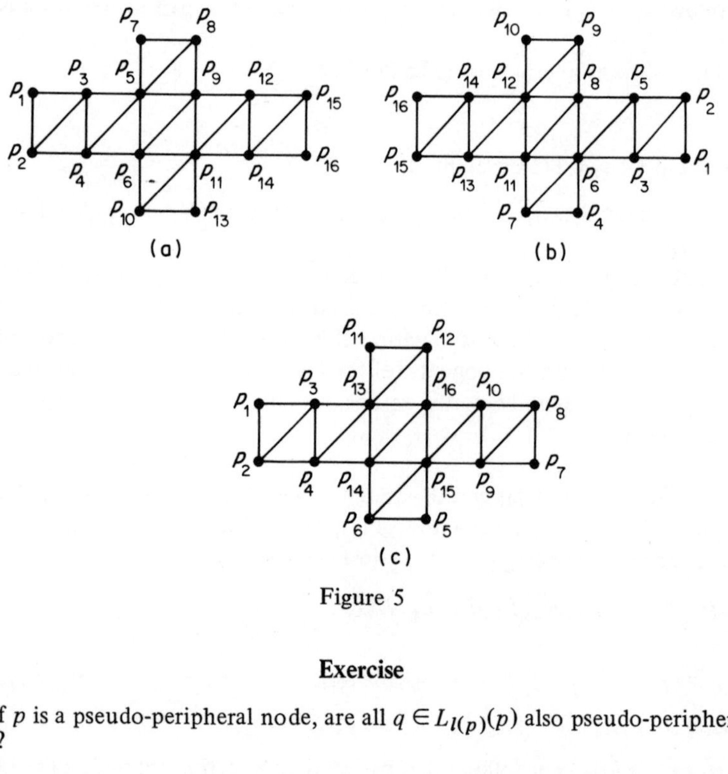

Figure 5

Exercise

4.4. If p is a pseudo-peripheral node, are all $q \in L_{l(p)}(p)$ also pseudo-peripheral nodes?

(b) Nested dissection

The object of the method discussed in this section is to reduce the storage required below that needed using a reverse Cuthill–McKee ordering. The price paid for this reduction will be a significant increase in the complexity of the program.

Example Let the nodes of the grid used in the previous example be numbered as in Figure 5(c). That is, number the branches, then number the centre. Then the matrix has the form

$$A = \begin{bmatrix} A_{11} & & & & \\ & A_{22} & & & \\ & & A_{33} & & \\ & & & A_{44} & \\ A_{15}^{T} & A_{25}^{T} & A_{35}^{T} & A_{45}^{T} & A_{55} \end{bmatrix}. \tag{4.9}$$

The form is known as *block diagonal and bordered*, but more important is the structure of the blocks:

A_{11}, A_{33}, A_{55} (4 x 4) dense

A_{22}, A_{44} (2 x 2) dense

while

A_{51}, A_{53} (4 x 4) sparse

A_{52}, A_{54} (4 x 2) sparse.

Clearly in this small example the distinction between a sparse block and a dense block is not very great, but it is still possible to illustrate the general approach. If it is possible to store only the lower triangles of the diagonal blocks, then (ignoring pointers and programming overheads) the storage required is 10 + 3 + 10 + 3 + 10 = 36 compared to 50 for A_{RCM}. In engineering terms it is possible to view the sets of nodes of A_{11}, A_{22}, A_{33} and A_{44} as *substructures* connected by the nodes of A_{55}.

If

$$A = LL^{T},$$

then L has the same block diagonal bordered form

$$L = \begin{bmatrix} L_{11} & & & & \\ & L_{22} & & & \\ & & L_{33} & & \\ & & & L_{44} & \\ W_{15}^{T} & W_{25}^{T} & W_{35}^{T} & W_{45}^{T} & L_{55} \end{bmatrix}, \tag{4.10}$$

but all the blocks will in general be dense, unlike the original matrix, in which the off-diagonal blocks are sparse. The reason for the transpose notation for the last row will become clear in the subsequent algebra.

It follows from (4.9) and (4.10) that

$$L_{ii}L_{ii}^{T} = A_{ii}, \qquad i = 1, 2, 3, 4, \tag{4.11}$$

$$L_{ii}W_{i5} = A_{i5} \qquad i = 1, 2, 3, 4 \tag{4.12}$$

and

$$L_{55}L_{55}^{\mathrm{T}} = A_{55} - \tilde{A}_{55}(\equiv \hat{A}_{55}), \tag{4.1}$$

where

$$\tilde{A}_{55} \equiv \sum_{i=1}^{4} W_{i5}^{\mathrm{T}} W_{i5}. \tag{4.1}$$

The method adopted for the factorization is then to compute L_{ii} $(i = 1, 2, 3, 4)$ from (4.11) by Cholesky factorization and store them in place of A_{ii} $(i = 1, 2, 3, 4)$. The matrix $\tilde{A}_{55}$ is then constructed without storing any of matrices W_{i5}. The matrix $\tilde{A}_{55}$ is constructed a column at a time and hence from (4.13) $\hat{A}_{55}$ is constructed column by column so that storage space is restricted. Only the diagon blocks need be stored in the computer core storage – the blocks A_{i5} are stored column by column on backing store.

The matrix $\tilde{A}_{55}$ is built up as components of

$$W_{i5}^{\mathrm{T}} W_{i5} \equiv A_{i5}^{\mathrm{T}}(L_{ii}^{-\mathrm{T}}(L_{ii}^{-1}A_{i5})). \tag{4.1}$$

If A_{i5} is stored column by column as

$$A_{i5} \equiv \left[\begin{array}{c|c|c} \mathbf{s}_1 & \cdots & \mathbf{s}_{n_5} \end{array}\right],$$

then for $i = 1, 2, 3, 4$ the jth column of (4.15) for $j = 1, \ldots, n_5$ is constructed by the following sequence:

(1) solve $L_{ii}\mathbf{t}_j = \mathbf{s}_j$, $\mathbf{t}_j$ is the jth column of W_{i5};
(2) solve $L_{ii}^{\mathrm{T}}\mathbf{u}_j = \mathbf{t}_j$, $\mathbf{u}_j$ is the jth column of $L_{ii}^{-\mathrm{T}} W_{i5}$;
(3) evaluate $\mathbf{v}_j = A_{i5}^{\mathrm{T}}\mathbf{t}_j$, $\mathbf{v}_j$ is the jth column of $W_{i5}^{\mathrm{T}} W_{i5}$;
(4) subtract $\mathbf{v}_j$ from the corresponding column of A_{55}.

Note that (1) and (2) only involve triangular matrices and so the solution is by forward substitution and backward substitution respectively. The vectors $\mathbf{s}_j$ and $\mathbf{t}_j$ are both of length n_i (the dimension of A_{ii}), whereas $\mathbf{v}_j$ is of length n_5. This form of implicit storage of the triangular factors is due to Jennings (1966).

The gains brought about by the block diagonal and bordered form of A may be further enhanced if the diagonal blocks are themselves block diagonal and bordered and so the implicit storage method outlined above can be used within each block. One algorithm that provides such a partitioning of the matrix is *nested dissection* (George, 1976).

Algorithm for nested dissection

Before describing the algorithm it is necessary to introduce two final concepts from graph theory. An undirected graph in which there is a path between every pair of nodes is called a *connected graph*, a disconnected graph is made up of *connected*

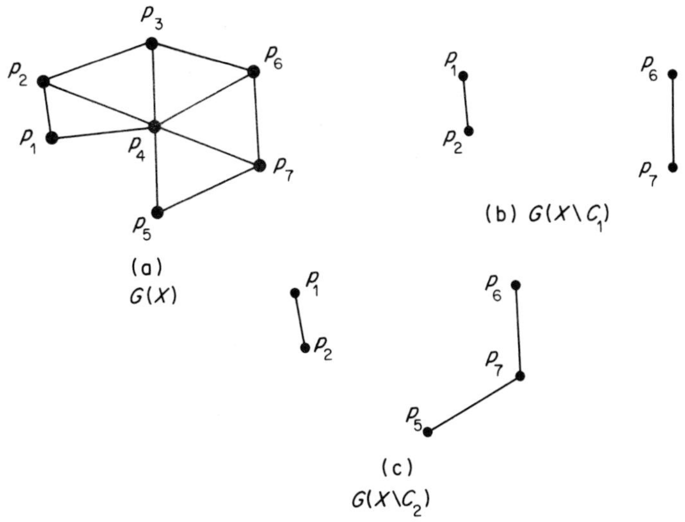

Figure 6

components. A set of nodes whose removal disconnects a previously connected component is called a *separator*. A *minimal separator* contains no proper subset that is a separator. An example of a minimal separator is illustrated in Figure 6, where

$$X \equiv \{p_1, p_2, p_3, p_4, p_5, p_6, p_7\}$$

$$C_1 \equiv \{p_3, p_4, p_5\}$$

and

$$C_2 \equiv \{p_3, p_4\}.$$

The graph $G(X)$ is connected while the graphs $G(X\backslash C_1)$ and $G(X\backslash C_2)$ are disconnected. Thus C_1 and C_2 are both separators and since $C_2 \subset C_1$, C_2 is a minimal separator and C_1 is not a minimal separator.

The algorithm can then be written:

(1) Initially $X = \{\text{all nodes}\}$ and $C = \emptyset$.

(2) Select a connected component of $G(X\backslash C)$ and find a pseudo-peripheral node $p \in X\backslash C$.

(3) Generate

$$\mathcal{L}(p) = \{L_0, \ldots, L_{l(p)}\}.$$

(4) If $l(p) \leqslant 1$, then the graph is not worth dissecting and $S = \mathcal{L}(p)$. Otherwise select a level (say L_j) near the middle of $\mathcal{L}(x)$. Choose $S \subset L_j$ that is a minimal separator of $G(X\backslash C)$.

(5) Number the nodes in S from $|X\backslash C|$ *downwards* in order of *decreasing* degree.

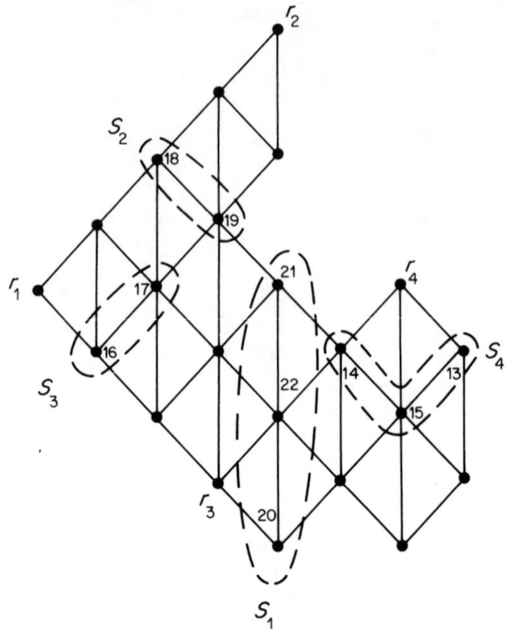

Figure 7

(6) Add the set S to the set C. If some nodes are still without numbers, return to (2).

Example In the grid given in Figure 7, the node marked r_1 is a pseudo-peripheral node, $l(r_1) = 7$ and $S_1 \subset L_4(r_1)$. The node r_2 is a pseudo-peripheral node of a connected component of $G(X \backslash S_1)$, $l(r_2) = 4$ and $S_2 \subset L_2(r_2)$. The nodes r_3 and r_4 are pseudo-peripheral nodes of their respective components of $G(X \backslash (S_1 \cup S_2))$ and they lead to separators S_3 and S_4 respectively. Thereafter no components warrant further dissection. The numbering within $S_1 \cup S_2 \cup S_3 \cup S_4$ is indicated. The order in which the remaining components are taken is not important, provided the nodes of an individual component are ordered consecutively.

Exercises

4.5. Prove that, in a level structure $\mathcal{L} = \{L_0, \ldots, L_l\}$ of X, if $S \subset L_j$ is the set of nodes connected to L_{j+1}, then S is a separator of $G(X)$. Under what conditions is such a set S a minimal separator?

4.6. Complete the ordering of the grid in Figure 7 and construct the corresponding matrix. Which blocks will be stored explicitly?

4.7. Construct the reverse Cuthill–McKee ordering of the grid in Figure 7.

PART II

Non-linear Equations

5

One-point Iteration Formulae

5.1 Introduction

These methods – alternatively called *functional iteration* – are the most straightforward to define and they include all the examples given in Section 1.2. The solution of an equation $f(x) = 0$ is obtained by computing a sequence $\{x^{(n)}\}$ $(n = 0, 1, \ldots)$ which, if the method is successful, converges to the solution α. The first approximation $x^{(0)}$ is given as part of the input data. Subsequent approximations are given by

$$x^{(n+1)} = \varphi(x^{(n)}), \qquad n = 0, 1, \ldots, \tag{5.1}$$

where the equation

$$x = \varphi(x)$$

has the same solution as $f(x) = 0$; thus

$$\alpha = \varphi(\alpha).$$

The formula (5.1) can be obtained from a simple rearrangement of terms; thus, for example, if

$$f(x) \equiv x^3 - 7x - 6,$$

it is possible to define a number of different *iteration functions* $\varphi(x)$ such as

$$\varphi(x) \equiv \tfrac{1}{7}(x^3 - 6) \tag{5.2}$$

or

$$\varphi(x) \equiv \frac{7x + 6}{x^2}. \tag{5.3}$$

Numerical example (simple iteration) Since

$$x^3 - 7x - 6 \equiv (x + 1)(x + 2)(x - 3),$$

the iterations with (5.2) and (5.3) were started with $x^{(0)} = -1.1$ and $x^{(0)} = -2.2$. The iteration (1.1) was also used and in this example (1.1) becomes

$$x^{(n+1)} = \frac{2x^{(n)3} + 6}{3x^{(n)2} - 7}.$$

The terminating condition in all cases was to stop the iteration if

$$|x^{(n+1)} - x^{(n)}| \leqslant 10^{-5} \qquad \text{(convergence)},$$
$$|x^{(n+1)}| \geqslant 10^{5} \qquad \text{(divergence)}$$

or

$$n = 20 \qquad \text{(too slow)}.$$

The results of the various iterations are given in Table 1, where a floating point format is used for large numbers, thus

$$-2.6\ \text{E} + 5 \equiv -260000.$$

Table 1

n	$\varphi(x)$: $\frac{1}{7}(x^3 - 6)$	$\varphi(x)$: $\frac{7x+6}{x^2}$	$\varphi(x)$: $\frac{2x^3+6}{3x^2-7}$
0	−1.10000	−1.10000	−1.10000
1	−1.04729	−1.40496	−0.99050
2	−1.02124	−1.94270	−0.99993
3	−1.00930	−2.01344	−1.00000
4	−1.00402	−1.99659	−1.00000
5	−1.00173	−2.00085	
6	−1.00074	−1.99979	
7	−1.00032	−2.00005	
8	−1.00014	−1.99999	
9	−1.00006	−2.00000	
10	−1.00003	−2.00000	
11	−1.00001		
12	−1.00000		
0	−2.20000	−2.20000	−2.20000
1	−2.37829	−1.94215	−2.03404
2	−2.77888	−2.01356	−2.00130
3	−3.92271	−1.99656	−2.00000
4	−9.48022	−2.00086	−2.00000
5	−12.25766	−1.99979	
6	−2.6 E + 5	−2.00005	
7		−1.99999	
8		−2.00000	
9		−2.00000	

Clearly, not all combinations of iteration formula and initial approximation are equally successful. In order to distinguish in advance between those iterations which will converge and those which will not, it is necessary to introduce some mathematics – the *contraction mapping theorem.*

5.2 Contraction Mapping Theorem

Theorem *Let φ be a continuous function that maps an N-dimensional closed and bounded region R into itself. That is, when* $\mathbf{x} \equiv [x_1, \ldots, x_N]^T$, *it follows that*

$$\varphi(\mathbf{x}) \equiv \begin{bmatrix} \varphi_1(x_1, \ldots, x_N) \\ \vdots \\ \varphi_N(x_1, \ldots, x_N) \end{bmatrix}$$

and that when $\mathbf{x} \in R$, *it also follows that* $\varphi(\mathbf{x}) \in R$. *Assume that there exists a positive constant* $L < 1$, *such that*

$$\| \varphi(\mathbf{a}) - \varphi(\mathbf{b}) \| \leqslant L \, \| \mathbf{a} - \mathbf{b} \| \tag{5.4}$$

for all $\mathbf{a}, \mathbf{b} \leqslant R$, *where* $\| \;\; \|$ *can be any norm. Then in R there is a unique solution of the equation* $\mathbf{x} = \varphi(\mathbf{x})$ *and the sequence* $\{\mathbf{x}^{(n)}\}$ $(n = 0, 1, \ldots)$ *such that*

$$\mathbf{x}^{(n+1)} = \varphi(\mathbf{x}^{(n)}), \qquad n = 0, 1, \ldots,$$

converges to this solution of $\mathbf{x} = \varphi(\mathbf{x})$ *for any initial approximation* $\mathbf{x}^{(0)} \in R$. Δ

Reminder A bounded region is one which has finite size. A closed region is one which includes its own boundary. In one dimension,

$[0, 1] \equiv \{x : 0 \leqslant x \leqslant 1\}$ is closed and bounded,

$(0, 1) \equiv \{x : 0 < x < 1\}$ is bounded but not closed

and

$(0, \infty) \equiv \{x : 0 < x\}$ is neither bounded nor closed.

A norm is a measure of size, and throughout this book particular examples are introduced as and when required.

This theorem clearly specifies conditions under which an iteration converges. These conditions, which are *sufficient but not necessary*, are that there exists a closed bounded region R in which the iteration function is a contraction mapping.

The use of the contraction mapping theorem

The behaviour of the iterations (5.2) and (5.3) as given in Table 1 is now explained using the contraction mapping theorem with the norm defined by

$\| x \| = | x |$. Consider first

$$\varphi(x) = \tfrac{1}{7}(x^3 - 6).$$

Since it is known from Table 1 that for some initial approximations the iteration converges to the root $x = -1$, it is possible to simplify the algebra by considering

$$x = -1 + \epsilon \qquad (\epsilon \text{ positive or negative}).$$

With this substitution,

$$\varphi(x) = \frac{-7 + 3\epsilon - 3\epsilon^2 + \epsilon^3}{7}$$

$$= -1 + \frac{\epsilon}{7}(3 - 3\epsilon + \epsilon^2),$$

if $| \epsilon | < 1$, it follows that

$$| \tfrac{1}{7}(3 - 3\epsilon + \epsilon^2) | < 1,$$

and so φ maps any interval $[-1 - \epsilon, -1 + \epsilon]$ $(0 < \epsilon < 1)$ into itself. In fact the same is true of each of the intervals $[-1 - \epsilon, -1]$ and $[-1, -1 + \epsilon]$. In order to complete the proof, it is necessary to know the value of the *Lipschitz constant L* that appears in (5.4). When the iteration function is continuously differentiable it is possible to determine the value of the constant using the *mean value theorem.*

The Mean Value Theorem (for functions of a single variable) *For any continuously differentiable function φ and any a, b,*

$$\varphi(a) - \varphi(b) = \varphi'(\xi)(a - b)$$

where $\xi = \theta a + (1 - \theta)b$ for some $\theta \in (0, 1)$. Δ

It follows from this theorem that for any $a, b \in R$

$$| \varphi(a) - \varphi(b) | \leqslant \max_{\xi \in R} | \varphi'(\xi) | \; | a - b |;$$

hence for functions of a single variable, it is possible to replace the Lipschitz constant in (5.4) by

$$\max_{\xi \in R} | \varphi'(\xi) |.$$

For the iteration function (5.2), it follows that

$$| \varphi'(\xi) | = \tfrac{3}{7}\xi^2,$$

which is bounded by unity if ξ lies in $(-\sqrt{\tfrac{7}{3}}, +\sqrt{\tfrac{7}{3}})$. Thus

$$\xi \in [-\sqrt{\tfrac{7}{3}} + \epsilon^*, +\sqrt{\tfrac{7}{3}} - \epsilon^*]$$

for some arbitrarily small positive number ϵ^*, and it follows that

$$| \max \varphi'(\xi) | = L < 1,$$

where the value of L clearly depends on the choice of ϵ^*. Thus the iteration will converge for any starting approximation in the interval

$$R = [-\sqrt{\tfrac{7}{3}} + \epsilon^*, +\sqrt{\tfrac{7}{3}} - \epsilon^*] \cap [-1 - \epsilon, -1 + \epsilon]$$
$$= [-\sqrt{\tfrac{7}{3}} + \epsilon^*, -1 + \epsilon], \tag{5.5}$$

where $0 < \epsilon < 1$ and ϵ^* is arbitrarily small and positive. This gives an interval that contains the point $x = -1.1$, whereas it does not contain the point $x = -2.2$.

For $\varphi(x)$ given by (5.3), the situation is somewhat different. Since

$$\varphi'(x) = -\frac{7x + 12}{x^3}$$

and extreme values of φ correspond to zeros of φ', it follows that the only turning point is at $x = -\frac{12}{7}$, at which $\varphi(x) = -\frac{49}{24}$. Therefore as

$$\varphi(-1) = \varphi(-6) = -1,$$

it follows that in this case φ maps the interval $[-6, -1]$ onto the sub-interval $[-\frac{49}{24}, -1]$. In a similar manner it can be shown that φ' has a local minimum at $x = -\frac{18}{7}$ at which $\varphi' = -\frac{343}{972}$. The functions φ and φ' are illustrated in Figure 8

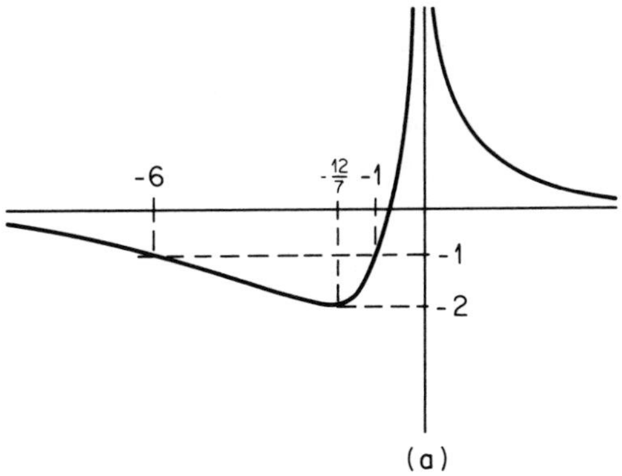

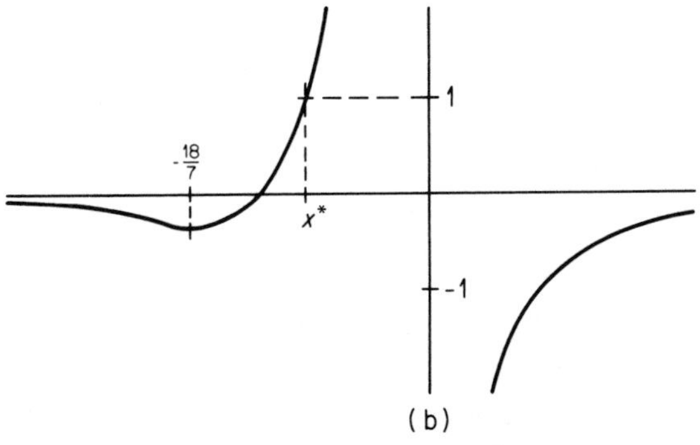

Figure 8 (a) $\varphi(x)$; (b) $\varphi'(x)$

Since $\varphi'(x^*) = 1$, where $x^* \triangleq -1.358$ and

$$\lim_{x\to\infty} \varphi'(x) = 0,$$

it follows that the iteration will converge for any initial approximation in

$$\begin{aligned} R &= (-\infty, x^* - \epsilon^*] \cap [-6, -1] \\ &= [-6, x^* - \epsilon^*] \end{aligned}$$

for arbitrarily small positive ϵ^*. This interval contains $x = -2.2$ but not $x = -1.1$ and therefore this result does not explain the convergence with the latter initial approximation (see Exercise 5.3). The third iteration illustrated in Table 1 is an example of *Newton's method*, which is considered in detail in Section 5.6. The reason for the superior convergence is also given later.

The contraction mapping theorem can be applied to systems of equations. For example in two dimensions $\mathbf{x} = [x_1, x_2]^T$,

$$\varphi(\mathbf{x}) = \begin{bmatrix} \varphi_1(x_1, x_2) \\ \varphi_2(x_1, x_2) \end{bmatrix}$$

and the iteration $\mathbf{x}^{(n+1)} = \varphi(\mathbf{x}^{(n)})$ is used to compute the solution of two simultaneous non-linear equations in the variables x_1 and x_2. In order to prove convergence in this case, it would be sufficient to find two closed bounded intervals R_1 and R_2 such that if $x_1 \in R_1$ and $x_2 \in R_2$ then $\varphi_1(x_1, x_2) \in R_1$ and $\varphi_2(x_1, x_2) \in R_2$, and in addition (5.4) is satisfied for some norm, the contraction mapping theorem is applied to systems in Section 5.4.

Exercises

5.1. Verify the region given in (5.5) by plotting the graphs of $\varphi(x)$ and $\varphi'(x)$.

5.2. Prove that (5.2) is a contraction mapping on each of the 'one-sided' intervals $[-\sqrt{\frac{7}{3}} + \epsilon^*, -1]$ and $[-1, -1 + \epsilon]$ with ϵ and ϵ^* as in (5.5). Do analogous 'one-sided' intervals exist for (5.3)?

5.3. Why does the iteration (5.3) converge when $x^{(0)} = -1.1 \notin R$? (*Hint*: What about $x^{(1)}$?)

5.4. A root of $x^3 - 3x^2 + 2x = 0$ is computed using the iteration formula

$$x^{(n+1)} = \frac{3x^{(n)2} - x^{(n)3}}{2}, \qquad n = 0, 1, \ldots$$

(a) verify that convergence to the solution $x = 0$ is guaranteed if $x^{(0)} \in [-\frac{1}{4}, \frac{1}{4}]$,
(b) find an interval in which convergence to the root $x = 2$ is guaranteed,
(c) prove that there is no interval in which convergence to the root $x = 1$ can be guaranteed. (*Hint*: What is the value of $\varphi'(1)$?)

5.5. Show that the iteration

$$x^{(n+1)} = (x^{(n)2} + x^{(n)} + 2)^{-1}, \qquad n = 0, 1, \ldots$$

will converge to a real root of

$$x^3 + x^2 + 2x - 1 = 0,$$

starting from *any positive* initial approximation to the root.

5.6. Determine whether the formulae;

(a) $$x^{(n+1)} = \tfrac{1}{4}\left(x^{(n)2} + \frac{6}{x^{(n)}}\right), \qquad n = 0, 1, \ldots$$

(b) $$x^{(n+1)} = 4 - \frac{6}{x^{(n)2}}, \qquad n = 0, 1, \ldots$$

are suitable to compute a root of the equation

$$x^3 = 4x^2 - 6$$

between 3 and 4.

5.3 Proof of the Contraction Mapping Theorem

In order to prove the theorem, it is necessary to use a preliminary lemma concerning the convergence of *Cauchy sequences.* A Cauchy sequence $\{\mathbf{a}^{(n)}\}$ $(n = 0, 1, \ldots)$ is such that for any arbitrarily small positive ϵ, there exists $N = N(\epsilon)$ such that for all $n, m \geqslant N$

$$\| \mathbf{a}^{(n)} - \mathbf{a}^{(m)} \| \leqslant \epsilon.$$

Lemma *A Cauchy sequence* $\{\mathbf{a}^{(n)}\}$ *of points from a closed bounded set* $R \subset \mathbb{R}^N$, *is a convergent sequence. That is, there exists* $\mathbf{a} \in R$ *such that*

$$\mathbf{a} = \lim_{n\to\infty} \mathbf{a}^{(n)}. \quad \Delta$$

Proof of the theorem For any $p > 0$, it follows from the formula $\mathbf{x}^{(n+1)} = \varphi(\mathbf{x}^{(n)})$, that

$$\| \mathbf{x}^{(p+1)} - \mathbf{x}^{(p)} \| = \| \varphi(\mathbf{x}^{(p)}) - \varphi(\mathbf{x}^{(p-1)}) \| \leqslant L \| \mathbf{x}^{(p)} - \mathbf{x}^{(p-1)} \|.$$

Replacing $\mathbf{x}^{(p)}$ by $\varphi(\mathbf{x}^{(p-1)})$ and so on eventually leads to

$$\| \mathbf{x}^{(p+1)} - \mathbf{x}^{(p)} \| \leqslant L^p \| \mathbf{x}^{(1)} - \mathbf{x}^{(0)} \|. \tag{5.6}$$

Also for any $p, q > 0$, repeated use of the triangle inequality:

$$\| \mathbf{x} - \mathbf{z} \| \leqslant \| \mathbf{x} - \mathbf{y} \| + \| \mathbf{y} - \mathbf{z} \|,$$

leads to

$$\| \mathbf{x}^{(p+q)} - \mathbf{x}^{(p)} \| \leqslant \| \mathbf{x}^{(p+q)} - \mathbf{x}^{(p+q-1)} \| + \| \mathbf{x}^{(p+q-1)} - \mathbf{x}^{(p+q-2)} \| + \cdots + \| \mathbf{x}^{(p+1)} - \mathbf{x}^{(p)} \|.$$

By combining inequalities, it follows that

$$\| \mathbf{x}^{(p+q)} - \mathbf{x}^{(p)} \| \leqslant \{L^{p+q-1} + L^{p+q-2} + \cdots + L^{p}\} \, \| \mathbf{x}^{(1)} - \mathbf{x}^{(0)} \|,$$

but

$$1 + L + L^2 + \cdots + L^{q-1} = \frac{1 - L^q}{1 - L} \leqslant \frac{1}{1 - L}$$

and so

$$\| \mathbf{x}^{(p+q)} - \mathbf{x}^{(p)} \| \leqslant \frac{L^p}{1 - L} \| \mathbf{x}^{(1)} - \mathbf{x}^{(0)} \|.$$

Thus for any $\epsilon > 0$, select N such that

$$\frac{L^N}{1 - L} \| \mathbf{x}^{(1)} - \mathbf{x}^{(0)} \| \leqslant \epsilon$$

and hence it follows that $\{\mathbf{x}^{(n)}\}$ is a Cauchy sequence. Then since R is closed and bounded, there exists $\alpha \in R$ such that

$$\alpha = \lim_{n \to \infty} \mathbf{x}^{(n)}.$$

Since $\varphi(\mathbf{x})$ is a continuous function

$$\lim_{n \to \infty} \varphi(\mathbf{x}^{(n)}) = \varphi(\alpha)$$

so

$$\alpha = \varphi(\alpha)$$

and thus the convergence of the iteration to a solution of the equation $\mathbf{x} = \varphi(\mathbf{x})$ is proved. Δ

For the one dimensional case the convergence or otherwise of an iteration $x^{(n+1)} = \varphi(x^{(n)})$ can be displayed as a path between the graphs $y = x$ and $y = \varphi(x)$ as in Figure 9.

Exercises

5.7. Prove by contradiction, that under the assumptions of the theorem, the solution of $\mathbf{x} = \varphi(\mathbf{x})$ is unique in R.

5.8. Prove that if $\mathbf{x}^{(n+1)} = \varphi(\mathbf{x}^{(n)})$ $(n = 0, 1, \ldots)$, then for $n = 1, 2, \ldots$

$$\| \alpha - \mathbf{x}^{(n)} \| \leqslant \frac{L}{1 - L} \| \mathbf{x}^{(n)} - \mathbf{x}^{(n-1)} \|. \qquad (5.7$$

5.9. Prove that if

$$x^{(n+1)} = \varphi(x^{(n)}) + e^{(n+1)}, \quad n = 0, 1, \ldots,$$

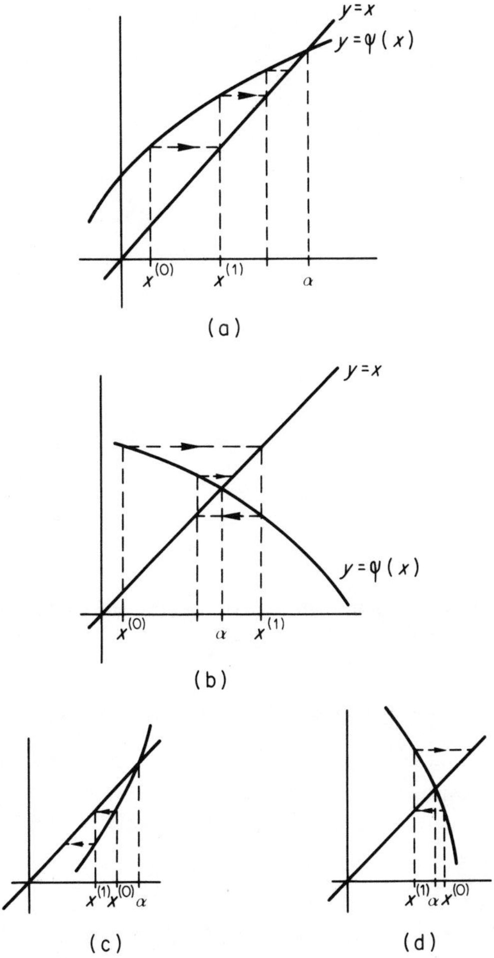

Figure 9 (a) $0 < \varphi'(x) < 1$;
(b) $-1 < \varphi'(x) < 0$; (c) $\varphi'(x) > 1$;
(d) $-1 < \varphi'(x)$

where $e^{(n)}$ could be round-off error introduced into the calculation, then

$$\| \alpha - x^{(n)} \| \leqslant \frac{1}{1-L} \{L \| x^{(n)} - x^{(n-1)} \| + \| e^{(n)} \|\}, \qquad n = 1, 2, \ldots .$$

5.10. Let $\{x^{(n)}\}$ be a sequence derived from the iteration

$$x^{(n+1)} = \varphi(x^{(n)}), \qquad n = 0, 1, \ldots,$$

where $\alpha = \varphi(\alpha)$ and $x^{(0)}$ is near to α such that $0 < \varphi'(x) \leqslant L < 1$ for all $\mathbf{x}$ in the interval defined by $x^{(0)}$ and α. Show that the sequence $\{x^{(n)}\}$ converges *monotonically* to α and that the sequence $\{y^{(n)}\}$ such that

$$y^{(n)} = \frac{x^{(n+1)} - x^{(n)}K}{1 - K} \qquad n = 0, 1, \ldots,$$

converges monotonically to α from the opposite side to the sequence $x^{(n)}$, provided that $0 < L < K < 1$.

5.11. Prove that if $x^{(0)}$ is sufficiently close to -1, the iteration defined by (5.2) always converges monotonically.

5.12. Show that if $\beta > 4\left(\frac{3}{4e}\right)^{3/4}$, the iteration

$$x^{(n+1)} = \frac{1}{\beta} e^{-x^{(n)4}} \qquad n = 0, 1, \ldots$$

will converge for any $x^{(0)} \geqslant 0$. Is the convergence monotonic? What would happen if $x^{(0)} < 0$?

5.13. The equation $3xe^{-x} = 1$ has a root near $x^{(0)} = 1.51$. Show that the iteration

$$x^{(n+1)} = \ln (3x^{(n)}), \qquad n = 0, 1, \ldots,$$

will converge, and using (5.6) and (5.7), estimate the least value of n for which $x^{(n)}$ is correct to six decimal places.

5.4 Systems of Non-linear Equations

Since the contraction mapping theorem is written in terms of N-dimensional space, there is no question of its not being valid for systems of equations. The only difficulty – apart from the increasing complexity of the algebra – is to determine the Lipschitz constant.

The Mean Value Theorem (for functions of several variables) *Let f be a single-valued function of N variables that is continuously differentiable with respect to each. Then for any* **a**, **b** $\in \mathbb{R}^N$, *there exists* $\theta \in [0, 1]$ *such that*

$$f(\mathbf{a}) - f(\mathbf{b}) = \nabla f(\boldsymbol{\xi})^{\mathrm{T}}(\mathbf{a} - \mathbf{b})$$

where $\boldsymbol{\xi} \equiv \theta\,\mathbf{a} + (1 - \theta)\mathbf{b}$ *and the gradient vector* ∇f *is defined as*

$$\nabla f = \left[\frac{\partial f}{\partial x_1}, \ldots, \frac{\partial f}{\partial x_N}\right]^{\mathrm{T}}. \qquad \Delta$$

In order to apply this theorem to each component of the *vector function* $\varphi(\mathbf{x})$, it is necessary to define $\theta_i \in [0, 1]$ $(i = 1, \ldots, N)$, and hence $\boldsymbol{\xi}_i = \theta_i \mathbf{a} + (1 - \theta_i)\mathbf{b}$ such that

$$\varphi(\mathbf{a}) - \varphi(\mathbf{b}) = J(\boldsymbol{\xi}_1, \ldots, \boldsymbol{\xi}_N)(\mathbf{a} - \mathbf{b}),$$

where the *matrix J* is defined as

$$J(\xi_1,\ldots,\xi_N) \equiv \begin{bmatrix} \dfrac{\partial \varphi_1(\xi_1)}{\partial x_1} & \cdots & \dfrac{\partial \varphi_1(\xi_1)}{\partial x_N} \\ \vdots & & \vdots \\ \dfrac{\partial \varphi_N(\xi_N)}{\partial x_1} & \cdots & \dfrac{\partial \varphi_N(\xi_N)}{\partial x_N} \end{bmatrix}.$$

Thus it is possible to apply the contraction mapping theorem to non-linear systems of equations if the Lipschitz constant is taken to be

$$L = \max_{\xi_1,\ldots,\xi_N \in R} \| J(\xi_1,\ldots,\xi_N) \|.$$

The matrix norm used in this definition is the norm subordinate to the vector norm appearing in (5.4); that is

$$\| J \| = \sup_{\|\mathbf{x}\|=1} \| J\mathbf{x} \|.$$

It is possible to prove a stronger result (see for example Goldstein, 1967) that

$$\| \varphi(\mathbf{a}) - \varphi(\mathbf{b}) \| \leqslant \max_{\xi \in R} \| J(\xi) \| \, \| \mathbf{a} - \mathbf{b} \|$$

where $J(\xi) \equiv J(\xi, \xi, \ldots, \xi)$ is called the *Jacobian matrix.*

Numerical example The contraction mapping theorem is used to investigate the convergence of the iteration

$$\mathbf{x}^{(n+1)} = \mathbf{x}^{(n)} - \begin{bmatrix} -\frac{1}{4} & & \\ & \frac{1}{5} & \\ & & -\frac{1}{2} \end{bmatrix} f(\mathbf{x}^{(n)}), \qquad n = 0, 1, \ldots, \tag{5.8a}$$

to the solution of

$$\mathbf{f}(\mathbf{x}) \equiv \begin{bmatrix} x_1^2 + x_2^2 - x_3 - 2 \\ x_1 + 5x_2 + 1 \\ x_1 x_3 - 2x_1 + k \end{bmatrix} = 0, \tag{5.8b}$$

where $\mathbf{x} \equiv [x_1, x_2, x_3]^{\mathrm{T}}$ and k is a constant parameter. General iterations of the form

$$\mathbf{x}^{(n+1)} = \mathbf{x}^{(n)} - a\mathbf{f}(\mathbf{x}^{(n)}), \qquad a = \text{constant},$$

are sometimes referred to as *Whittaker's method* (Whittaker, 1918). To apply the contraction mapping theorem in this problem, it is necessary to find a closed bounded region $R \subset \mathbb{R}^3$ such that $\varphi(\mathbf{x}) \in R$ whenever $\mathbf{x} \in R$ and in which

$$\| \varphi(\mathbf{x}) - \varphi(\mathbf{y}) \| \leqslant L \| \mathbf{x} - \mathbf{y} \|$$

in some norm. The Jacobian matrix of $\varphi(\mathbf{x})$ is

$$J_\varphi(\mathbf{x}) = \begin{bmatrix} 1+\frac{1}{2}x_1 & \frac{1}{2}x_2 & -\frac{1}{4} \\ -\frac{1}{5} & 0 & 0 \\ \frac{1}{2}x_3 - 1 & 0 & \frac{1}{2}x_1 + 1 \end{bmatrix},$$

and so it is sensible to search for positive constants σ_1, σ_2 and σ_3 such that $\varphi(\mathbf{x})$ is a contraction mapping on the region

$$R = \{\mathbf{x} : |2 + x_1| \leqslant \sigma_1, |x_2| \leqslant \sigma_2, |2 - x_3| \leqslant \sigma_3\}.$$

To simplify the algebra, define $\epsilon_1 \equiv 2 + x_1$ and $\epsilon_3 \equiv x_3 - 2$. If $\varphi(\mathbf{x})$ is a mapping of R into R it follows that

$$|\varphi_1(\mathbf{x}) + 2| \leqslant \sigma_1,$$

$$|\varphi_2(\mathbf{x})| \leqslant \sigma_2$$

and

$$|\varphi_3(\mathbf{x}) - 2| \leqslant \sigma_3.$$

It is a contraction mapping if, in addition, it follows that

$$\max_{\mathbf{x}\in R} \|J(\mathbf{x})\| \equiv \max_{\substack{|\epsilon_1|\leqslant\sigma_1, |x_2|\leqslant\sigma_2,\\ |\epsilon_3|\leqslant\sigma_3}} \left\| \begin{bmatrix} \frac{1}{2}\epsilon_1 & \frac{1}{2}x_2 & -\frac{1}{4} \\ -\frac{1}{5} & 0 & 0 \\ \frac{1}{2}\epsilon_3 & 0 & \frac{1}{2}\epsilon_1 \end{bmatrix} \right\| \leqslant L < 1.$$

If the l_∞-norm is used, it follows that this latter condition is equivalent to the condition

$$\max_{\sigma_1,\sigma_2,\sigma_3} \{\tfrac{1}{2}\sigma_1 + \tfrac{1}{2}\sigma_2 + \tfrac{1}{4}, \tfrac{1}{5}, \tfrac{1}{2}\sigma_3 + \tfrac{1}{2}\sigma_1\} \leqslant L.$$

Thus a set of conditions that are *sufficient* to ensure convergence, consists of the inequalities

$$\sigma_1 + \sigma_2 \leqslant 2L - \tfrac{1}{2},$$

$$\sigma_1 + \sigma_3 \leqslant 2L,$$

$$\left.\begin{aligned} &\tfrac{1}{4}|\epsilon_1^2 + x_2^2 - \epsilon_3| \leqslant \sigma_1, \\ &\tfrac{1}{5}|1 - \epsilon_1| \leqslant \sigma_2 \\ \text{and}\quad & \\ &\tfrac{1}{2}|k - \epsilon_1\epsilon_3| \leqslant \sigma_3 \end{aligned}\right\} \quad |\epsilon_1| \leqslant \sigma_1, |x_2| \leqslant \sigma_2, |\epsilon_3| \leqslant \sigma_3.$$

One solution when $k \in [-1, 1]$, though not necessarily the one leading to the largest region R, is $\sigma_1 = \sigma_2 = \frac{1}{4}$, $\sigma_3 = \frac{3}{4}$. From this solution it follows that a realistic value for the Lipschitz constant in R is $L = \frac{1}{2}$. Two examples of the iteration (5.8a) with $\mathbf{x}^{(0)} = [-2, 0, 2]^{\mathrm{T}}$ and $k = \pm 1$ are given in Table 2, and it can be seen from the

Table 2 $\mathbf{x}^{(0)} = [-2, 0, 2]^T$

(a) $k = 1$

n	x_1	x_2	x_3
1	−2.000000	0.200000	2.500000
2	−2.115000	0.223000	2.471250
3	−2.102074	0.220415	2.475945
4	−2.104237	0.220847	2.475194
5	−2.103889	0.220778	2.475316
6	−2.103945	0.220789	2.475297
7	−2.103936	0.220787	2.475300
8	−2.103938	0.220787	2.375299
9	−2.103937	0.220787	2.475299
10	−2.103937	0.220787	2.475299

(b) $k = -1$

n	x_1	x_2	x_3
1	−2.000000	0.200000	1.500000
2	−1.865000	0.173000	1.466250
3	−1.854524	0.170904	1.461176
4	−1.852701	0.170540	1.460316
5	−1.852384	0.170477	1.460167
6	−1.852329	0.170466	1.460141
7	−1.852317	0.170464	1.460137
8	−1.852317	0.170463	1.460135
9	−1.852317	0.170463	1.460136

results that the Lipschitz constant leads a very poor estimate of the errors when (5.7) is applied to this example, for $L/(1 - L) = 1$, whereas

$$\| \mathbf{x}^{(n)} - \boldsymbol{\alpha} \|_\infty \ll \| \mathbf{x}^{(n)} - \mathbf{x}^{(n-1)} \|_\infty, \qquad n = 1, 2, \ldots .$$

Methods for solving non-linear equations and methods for finding the maxima and minima of non-linear functions are closely linked in practice, since the two problems are closely linked in theory. A necessary and sufficient condition for a smooth function $F(\mathbf{x})$ to have a stationary point at $\mathbf{x} = \mathbf{x}^*$ is that

$$\nabla F(\mathbf{x}^*) = 0.$$

That is the function $\mathbf{f} \equiv \nabla F$ has a zero at $\mathbf{x} = \mathbf{x}^*$. In such problems it follows that

$$\frac{\partial f_i}{\partial x_j} \equiv \frac{\partial^2 F}{\partial x_i \partial x_j} \equiv \frac{\partial f_j}{\partial x_i},$$

so the Jacobian matrix is symmetric and hence special methods for this case are of particular importance. There is a considerable and growing literature on this subject (for example Lootsma, 1972; Murray, 1972; Jacobs, 1977) and in Section 7.4 we return to the problem of finding minima in the particular case when

$$F(\mathbf{x}) \equiv \mathbf{f}(\mathbf{x})^T \mathbf{f}(\mathbf{x}).$$

In this special situation the function to be minimized is the l_2-norm of the residual and so since **f** will be in general a non-linear function, this is a *non-linear least squares problem* (cf. Section 3.3).

Exercises

5.14. Let A be a non-singular square matrix and let $G^{(0)}$ be an approximation to A^{-1}. Show that if the iteration defined by

$$\begin{aligned} R^{(n)} &= I - AG^{(n)}, \\ G^{(n+1)} &= G^{(n)} + G^{(n)}R^{(n)} \end{aligned} \qquad n = 0, 1, \ldots$$

converges then $\lim_{n\to\infty} G^{(n)} = A^{-1}$. Explain the connexion with Newton's method for

$$f(x) \equiv \frac{1}{x} - a.$$

5.15. Replace the l_∞-norm by the l_1-norm and show that it is still possible to verify that $\varphi(\mathbf{x})$ from (5.8a) and (5.8b) is a contraction mapping. Show that in this case it is necessary to take $L = \frac{5}{8}$.

5.16. Verify that the iteration

$$\mathbf{x}^{(n+1)} = \mathbf{x}^{(n)} - \begin{bmatrix} & & -\frac{1}{4} \\ & +\frac{1}{5} & \\ -1 & & \end{bmatrix} \mathbf{f}(\mathbf{x}^{(n)}), \qquad n = 0, 1, \ldots$$

is suitable for computing a solution of (5.8b) near $\mathbf{x}^* = [0, 0, -2]^T$. In particular prove that the iteration will converge for any starting approximation in

$$R = \{\mathbf{x}\colon |x_1|, |x_2| \leqslant 0.3, |x_3 + 2| \leqslant 0.2\}$$

for any $k \in [-1, 1]$. Verify that $L = 0.85$ is a suitable value for the Lipschitz constant if the l_1-norm is used.

5.17. Let

$$R = \{\mathbf{x}\colon |x_1| \leqslant 0.2, x_2 \in [0.2, 0.4], x_3 \in [0.6, 0.9]\},$$

$$\mathbf{f}(\mathbf{x}) = \begin{bmatrix} x_1x_3 + x_2^3 - 0.1 \\ \frac{1}{3}x_1^3 + x_2^2 - 0.1 \\ x_3^2 - x_1x_2 - 0.5 \end{bmatrix}$$

and

$$A = \begin{bmatrix} 1 & & \\ & \frac{5}{3} & \\ & & \frac{2}{3} \end{bmatrix}.$$

Then verify that the iteration

$$\mathbf{x}^{(n+1)} = \mathbf{x}^{(n)} - A\mathbf{f}(\mathbf{x}^{(n)}), \qquad n = 0, 1, \ldots$$

will converge for any $\mathbf{x}^{(0)} \in R$.

5.5 Iterative Methods for Systems of Linear Equations

Iterative methods that are used in practice tend to be tailored to the particular form of the equations – they are usually formed by replacing derivatives in partial differential equations by differences. Any comparison of such methods involves aspects of matrix theory which are beyond the scope of this volume (see, for example, Varga, 1962 or Ortega 1972). Iterative methods are of importance because practical problems invariably lead to large sparse matrices which are highly structured (a few examples appear in Section 4.3) and iterative methods can make good use of these properties to provide highly efficient algorithms. In general, elementary textbooks discuss only the simplest methods, namely Jacobi and Gauss–Seidel. These are not of great practical significance in many problems. For completeness this section, concerned with the basic properties of iterative methods for linear systems, includes a description of these two methods, but for a more detailed discussion the reader is directed elsewhere (for example Forsythe and Moler, 1967).

The majority of iterative methods for linear systems are *stationary linear iterations*; that is they can be written as

$$\mathbf{x}^{(n+1)} = M\mathbf{x}^{(n)} + \mathbf{g}, \qquad n = 0, 1, \ldots, \tag{5.9}$$

where M is a *constant* matrix and $\mathbf{g}$ is a constant vector. It follows from the contraction mapping theorem that a sufficient condition for the convergence of this iteration is that

$$\| M \| < 1,$$

for any norm.

Theorem *A necessary and sufficient condition for convergence from an arbitrary initial approximation is that*

$$\rho(M) < 1, \tag{5.10}$$

where $\rho(\ )$ *denotes the* spectral norm – *that is the absolute value of the eigenvalue with the maximum modulus.* △

Outline of proof The sufficiency follows directly from the contraction mapping theorem. The proof of necessity is not so straightforward. It follows from (5.9) that

$$\begin{aligned}\mathbf{x}^{(n+1)} - \mathbf{x}^{(n)} &= M(\mathbf{x}^{(n)} - \mathbf{x}^{(n-1)}) \\ &= M^n(\mathbf{x}^{(1)} - \mathbf{x}^{(0)}), \qquad n = 0, 1 \ldots .\end{aligned}$$

Thus if the sequence $\{\mathbf{x}^{(n)}\}$ converges, it follows that

$$\lim_{n\to\infty} (\mathbf{x}^{(n+1)} - \mathbf{x}^{(n)}) = 0;$$

hence

$$\lim_{n\to\infty} M^n = 0. \tag{5.11}$$

Since (5.10) and (5.11) are equivalent (a detailed proof of this equivalence is given in, for example, Wendroff (1966) p. 156) it follows that (5.10) is a necessary as well as sufficient condition. Δ

Splitting methods

The iteration function

$$\varphi(\mathbf{x}) = M\mathbf{x} + \mathbf{g},$$

can be obtained from the original linear equations by rearranging the equations. If

$$A \equiv B - C \tag{5.12}$$

such that B is a non-singular matrix, then the systems

$$A\mathbf{x} = \mathbf{b}$$

and

$$\mathbf{x} = B^{-1}C\mathbf{x} + B^{-1}\mathbf{b}$$

have the same solution. It only remains to identify $B^{-1}C$ with M and $B^{-1}\mathbf{b}$ with $\mathbf{g}$ and the equations are in the desired form. Any partition of A given by (5.12) is called a *splitting*.

For any matrix $A = \{a_{ij}\}$, define the matrices D, L and U such that

$$D = \operatorname{diag}(a_{ii}), \qquad \text{(diagonal)}$$

$$L = \begin{cases} -a_{ij}, & i > j \\ 0 & i \leqslant j \end{cases} \qquad \text{(strictly lower triangular)}$$

and

$$U = \begin{cases} -a_{ij}, & i < j \\ 0 & i \geqslant j, \end{cases} \qquad \text{(strictly upper triangular).}$$

Thus

$$A = D - L - U.$$

The (point) *Jacobi method* can then be written as

$$D\,\mathbf{x}^{(n+1)} = (L + U)\,\mathbf{x}^{(n)} + \mathbf{b}, \tag{5.13a}$$

or

$$a_{ii}x_i^{(n+1)} = -\sum_{\substack{j \neq i \\ j=1}}^{N} a_{ij}x_j^{(n)} + b_i, \qquad i = 1, 2, \ldots, N, \tag{5.13b}$$

whereas the (point) *Gauss–Seidel method* is

$$(D - L)\,\mathbf{x}^{(n+1)} = U\mathbf{x}^{(n)} + \mathbf{b} \tag{5.14a}$$

or

$$a_{ii}x_i^{(n+1)} = -\sum_{j=1}^{i-1} a_{ij}x_j^{(n+1)} - \sum_{j=i+1}^{N} a_{ij}x_j^{(n)} + b_i, \qquad i = 1, 2, \ldots, N. \tag{5.14b}$$

A necessary condition for these methods to be well defined is clearly that $a_{ii} \neq 0$ ($i = 1, \ldots, N$).

Although it would appear from (5.13b) and (5.14b) that Gauss–Seidel is the more complex algorithm, this is not in fact true. A standard method of implementing the Jacobi iteration is to evaluate the *residual vector*

$$\mathbf{r}(\mathbf{x}) \equiv \mathbf{b} - A\mathbf{x}$$

at each stage. If $\mathbf{r}^{(n)} \equiv \mathbf{r}(\mathbf{x}^{(n)})$, then (5.13b) can be written as

$$a_{ii}x_i^{(n+1)} = a_{ii}x_i^{(n)} + r_i^{(n)}, \qquad i = 1, 2, \ldots, N. \tag{5.15}$$

Hence, although it is possible to overwrite $\mathbf{x}^{(n)}$ with $\mathbf{x}^{(n+1)}$ it is necessary to retain the additional vector $\mathbf{r}^{(n)}$, and this may prove to be a handicap in a large problem. The residual vector is invariably used in the terminating condition, say

$$\| \mathbf{r}^{(n)} \|_1 < \epsilon,$$

for example; but as A^{-1} is unknown it is not possible to say very much about the ize of the *error* $\boldsymbol{\alpha} - \mathbf{x}^{(n)}$, even when the *residual* $\mathbf{r}^{(n)}(=A\boldsymbol{\alpha} - A\mathbf{x}^{(n)})$ is small (cf. iterative refinement in Section 2.4).

One method of implementing the Gauss–Seidel iteration is to use (5.15) but to define

$$r_i^{(n)} = b_i - \sum_{j=1}^{i-1} a_{ij}x_j^{(n+1)} - \sum_{j=i}^{N} a_{ij}x_j^{(n)}. \tag{5.16}$$

It is then not necessary to store all $r_i^{(n)}$ ($i = 1, \ldots, N$) simultaneously; each is omputed using the most recent information as the components of $\mathbf{x}$ are succes-ively updated and overwritten – hence the alternative name *successive displacements*. An alternative name for the Jacobi iteration is *simultaneous displacements*. As the Gauss–Seidel iteration updates the vector $\mathbf{x}$ using more recent information, it might be assumed that it converges faster than Jacobi. This is frequently true and if $a_{ij} < 0$ $(i \neq j)$, we have:

The Stein–Rosenburg theorem *If $a_{ij} < 0$ $(i \neq j)$, then either both the Jacobi and Gauss–Seidel iterations converge or neither converges, and if they converge, Gauss–Seidel converges faster than Jacobi.* △

The condition $a_{ij} < 0$ $(i \neq j)$ is not unreasonable in many applications, but when it is not satisfied, examples can be constructed such that Jacobi converges and Gauss–Seidel does not. A modification of the Gauss–Seidel method which frequently leads to a significant improvement is an iteration of the form

$$a_{ii}x_i^{(n+1)} = a_{ii}x_i^{(n)} + \omega r_i^{(n)}, \qquad i = 1, \ldots, N, \tag{5.17}$$

where $r_i^{(n)}$ is given by (5.16). With the parameter $\omega \in (1, 2)$ this is known as *successive-over-relaxation* (s.o.r.). The speed of convergence depends on the choice of ω and it is often difficult to estimate the optimum value. Investigations of the properties of s.o.r. have been the subject of considerable study (Young, 1971) and it is still in progress.

The Jacobi and Gauss–Seidel iterations are referred to as *point* methods to distinguish them from *block* methods. Block iteration is another modification of the basic method that can lead to a marked improvement in efficiency for many applications.

It is not uncommon in the solution of partial differential equations to derive *difference equations* that can be written in terms of *block matrices* (see Figure 18 in Section 7.3(b)) where the *diagonal blocks* are tridiagonal matrices (see Mitchell, 1969); thus

$$A \equiv \begin{bmatrix} A_{11} & A_{12} & \cdots & A_{1p} \\ \vdots & & & \vdots \\ A_{p1} & A_{p2} & \cdots & A_{pp} \end{bmatrix},$$

where

$$A_{ii} \equiv \begin{bmatrix} \alpha_1^{(i)} & \gamma_1^{(i)} & & \\ \beta_1^{(i)} & \ddots & \ddots & \gamma_{q-1}^{(i)} \\ & \ddots & \beta_{q-1}^{(i)} & \alpha_q^{(i)} \end{bmatrix}, \qquad i = 1, \ldots, p.$$

In addition, it is frequently the case that $A_{ij} \equiv 0 \mid i - j \mid > 1$. The A is a $pq \times pq$ matrix and if $\mathbf{x}$ and $\mathbf{b}$ are similarly partitioned into subvectors it follows that the system

$$A\mathbf{x} = \mathbf{b}$$

can be written as

$$\sum_{j=1}^{q} A_{ij}\mathbf{x}_j = \mathbf{b}_i, \qquad i = 1, \ldots, p.$$

The *block s.o.r.* iteration for such a system is

$$A_{ii}\mathbf{x}_i^{(n+1)} = A_{ii}\mathbf{x}_i^{(n)} + \omega\mathbf{r}_i^{(n)}, \qquad i = 1, \ldots, p, \tag{5.18}$$

where

$$\mathbf{r}_i^{(n)} = \mathbf{b}_i - \sum_{j=1}^{i-1} A_{ij}\mathbf{x}_j^{(n+1)} - \sum_{j=i}^{p} A_{ij}\mathbf{x}_j^{(n)}.$$

The formula (5.18) consists of p sets of $q \times q$ tridiagonal equations which – assuming that A_{ii} ($i = 1, \ldots, p$) are non-singular – can be solved by an efficient *direct method* (see Section 4.1).

Exercises

5.18. Verify that the (point) s.o.r. formula can be written as

$$\mathbf{x}^{(n+1)} = (D - wL)^{-1}([1 - \omega]D + \omega U)\,\mathbf{x}^{(n)} + \omega(D - \omega L)^{-1}\mathbf{b}.$$

5.19. Verify that both Jacobi and Gauss–Seidel iterations converge if A is diagonally dominant.

5.20. The matrix A is strictly diagonally dominant: that is

$$\sum_{\substack{j=1\\ j\neq i}}^{N} |a_{ij}| \leqslant \gamma < 1, \qquad i = 1, 2, \ldots, N.$$

The following iterative scheme is used to solve the set of equations $A\mathbf{x} = \mathbf{b}$:

$$x_1^{(k+1)} = b_1 - \sum_{j=2}^{n} a_{ij}x_j^{(k)},$$

$$x_2^{(k+1)} = b_2 - a_{21}\{\alpha x_1^{(k+1)} + (1-\alpha)x_i^{(k)}\} - \sum_{j=3}^{n} a_{2j}x_j^{(k)},$$

$$x_r^{(k+1)} = b_r - \sum_{j=1}^{r-1} a_{rj}\{\alpha x_j^{(k+1)} + (1-\alpha)x_j^{(k)}\}$$

$$- \sum_{j=r+1}^{n} a_{rj}x_j^{(k)}, \qquad k = 0, 1, \ldots.$$

Derive a set of sufficient conditions under which convergence of this scheme is assured.

5.21. Let

$$\mathbf{f}(\mathbf{x}) \equiv A\mathbf{x} - \mathbf{b} = 0$$

be a linear system of equations, where A is an $N \times N$ symmetric matrix with the eigenvalues

$$0 < \lambda_N \leqslant \lambda_{N-1} \leqslant \lambda_{N-2} \leqslant \cdots \leqslant \lambda_2 \leqslant \lambda_1.$$

For any initial vector $\mathbf{x}^{(0)}$, a sequence of successive approximate solutions $\{\mathbf{x}^{(n)}\}$ can be obtained from the iteration

$$\mathbf{x}^{(n+1)} = \mathbf{x}^{(n)} - a\mathbf{f}(\mathbf{x}^{(n)})$$

(Whittaker's method) for some real constant a. Write this iteration in the form

$$\mathbf{x}^{(n+1)} = M\mathbf{x}^{(n)} + \mathbf{g}$$

and show that $\rho(M)$ is a minimum when

$$a = \frac{2}{\lambda_1 + \lambda_N}.$$

5.22. Use the Jacobi and Gauss–Seidel iterations to solve the system

$$8x_1 + x_2 + x_3 = 9,$$
$$-x_1 + 7x_2 + 2x_3 = 7,$$
$$x_1 - x_2 + 10x_3 = 4.$$

Obtain answers correct to two decimal places using $\mathbf{x}^{(0)} = [1, 1, 0.4]^{\mathrm{T}}$.

5.6 Order of Convergence

Earlier in this chapter it is shown by the contraction mapping theorem that a sufficient condition for convergence of an iteration $x^{(n+1)} = \varphi(x^{(n)})$ is $|\varphi'| \leqslant L < 1$ in some neighbourhood of the solution. For systems of equations a sufficient condition is that $\| J(\xi) \| \leqslant L \leqslant 1$ for all ξ in some neighbourhood of the solution. It follows from the Lipschitz condition that

$$\| \alpha - x^{(n+1)} \| = \| \varphi(\alpha) - \varphi(x^{(n)}) \| \leqslant L \| \alpha - x^{(n)} \|,$$

and so the error should be reduced by a factor of (at least) L. Close to the solution, when φ' is nearly constant it follows that as $n \to \infty$

$$\alpha - x^{(n+1)} \approx \varphi'(\alpha)(\alpha - x^{(n)}).$$

For example in Table 1,

$$\varphi(x) = \tfrac{1}{7}(x^3 - 6)$$

leads to $\varphi'(-1) = \frac{3}{7} \triangleq 0.429$, and it is clear that

$$x^{(n+1)} + 1 \approx \tfrac{3}{7}(x^{(n)} + 1), \qquad n > 3.$$

Similarly

$$\varphi(x) = \frac{7x + 6}{x^2}$$

leads to $\varphi'(-2) = -\frac{1}{4}$ and for the corresponding iteration it is seen that

$$x^{(n+1)} + 2 \approx -\tfrac{1}{4}(x^{(n)} + 2) \qquad n > 3.$$

n the third iteration using Newton's method

$$x^{(n+1)} = x^{(n)} - \frac{f(x^{(n)})}{f'(x^{(n)})}, \qquad n = 0, 1, \ldots,$$

is not true that the error is reduced by a constant factor, as can be verified from able 3, where the results of the iteration are repeated correct to 7 decimal places.

able 3 Newton's method for $(x) \equiv x^3 - 7x - 6$

$x^{(n)}$	
−1.1	−2.2
−0.9905044	−2.0340426
−0.9999337	−2.0012994
−1.0000000	−2.0000020
−1.0000000	−2.0000000

The reason for the extremely rapid convergence of these iterations can be xplained using a *Taylor expansion* of the iteration function $\varphi(x)$ about the solution ; thus

$$x^{(n+1)} \equiv \varphi(x^{(n)}) = \varphi(\alpha) + \varphi'(\alpha)(x^{(n)} - \alpha) + \frac{\varphi''(\alpha)}{2}(x^{(n)} - \alpha)^2 + \cdots.$$

lence

$$x^{(n+1)} - \alpha = (x^{(n)} - \alpha)\left\{\varphi'(\alpha) + \frac{\varphi''(\alpha)}{2}(x^{(n)} - \alpha) + \cdots\right\}, \tag{5.19}$$

nd so when $\varphi'(\alpha) \neq 0$, it follows that

$$\lim_{x^{(n)} \to \alpha} \frac{x^{(n+1)} - \alpha}{x^{(n)} - \alpha} = \varphi'(\alpha);$$

hat is, the error is ultimately reduced by a constant factor and it will oscillate in ign if $0 > \varphi'(\alpha) > -1$. Newton's method – otherwise known as the Newton–Raphson method* – incorporates the iteration function

$$\varphi(x) \equiv x - \frac{f(x)}{f'(x)};$$

herefore

$$\varphi'(\alpha) = 1 + \frac{f(\alpha)f''(\alpha)}{[f'(\alpha)]^2} - \frac{f'(\alpha)}{f'(\alpha)}$$

See Historical Appendix

which is zero since $f(\alpha) = 0$, assuming $f'(\alpha) \neq 0$. In this case, it follows from (5.19) that

$$x^{(n+1)} - \alpha = (x^{(n)} - \alpha)^2 \left\{\frac{\varphi''(\alpha)}{2} + \frac{\varphi'''(\alpha)}{6}(x^{(n)} - \alpha) + \cdots\right\}$$

and so

$$\lim_{x^{(n)}\to\alpha} \frac{x^{(n+1)} - \alpha}{(x^{(n)} - \alpha)^2} = \frac{\varphi''(\alpha)}{2}.$$

Assuming that the factor $\varphi''(\alpha)/2$ is approximately unity, it follows that the number of decimal places correct is about doubled at each iteration.

Definition If the sequence $\{x^{(n)}\}$ tends to a limit α in such a way that

$$\lim_{x^{(n)}\to\alpha} \frac{x^{(n+1)} - \alpha}{(x^{(n)} - \alpha)^p} = C,$$

for some $C \neq 0$ and $p \geqslant 1$ then the order of convergence of the sequence is said to be p, and C is known as the *asymptotic error constant.*

Lemma The order of convergence of a convergent one-point iteration is p if and only if

$$\varphi'(\alpha) = \varphi''(\alpha) = \cdots = \varphi^{(p-1)}(\alpha) = 0$$

and $\varphi^{(p)}(\alpha) \neq 0$. When the order is p, the asymptotic error constant is

$$\frac{\varphi^{(p)}(\alpha)}{p!}. \quad \Delta$$

A proof follows directly from the Taylor expansion (5.19).

It is possible to define the order of a method for systems of equations, but it must be formulated in terms of

$$\lim_{\mathbf{x}^{(n)}\to\boldsymbol{\alpha}} \frac{\|\mathbf{x}^{(n+1)} - \boldsymbol{\alpha}\|}{\|\mathbf{x}^{(n)} - \boldsymbol{\alpha}\|^p}$$

and so the subsequent lemma is correspondingly more complex.

Exercises

5.23. Prove that for Newton's method, if $f'(\alpha) \neq 0$,

$$\frac{\varphi''(\alpha)}{2} = \frac{f''(\alpha)}{2f'(\alpha)}.$$

5.24. Show that if $|f''(\alpha)| < 4|f'(\alpha)|$ and $x^{(n)}$ is correct to p decimal places, then $x^{(n+1)}$ will be correct to $2p$ decimal places using Newton's method.

.25. With $x^{(0)}$ sufficiently close to $a^{1/2}$ and

$$\varphi_1(x) = \frac{x^2 + a}{2x} \qquad \text{and} \qquad \varphi_2(x) = \frac{x(3a - x^2)}{2a},$$

erify that

$$x^{(n+1)} = \varphi_i(x^{(n)}), \qquad i = 1, 2,$$

re both quadratically convergent, i.e. the order of convergence is 2. *Hence* prove hat

$$x^{(n+1)} = \varphi_3(x^{(n)}),$$

vhere

$$\varphi_3(x) = \frac{3(a + x^2)^2}{8ax} - \frac{x^3}{2a},$$

s of order 3.

.26. Under what conditions is Newton's method of order 3? The equation $e^{-x} - \sinh x = 0$ can be written as $e^{2x} = 3$; show that Newton's method is at least order 3 when applied to the first form but only order 2 for the second.

.27. Show that if the iteration formula

$$x^{(n+1)} = \frac{x^{(n)k} + kax^{(n)}}{kx^{(n)k-1} + a} \qquad n = 0, 1, \ldots$$

onverges, then the limit is the $(k - 1)$th root of a. Find the values of k for which he formula has an order of convergence $\geqslant 2$. Show that only one of these values of leads to a non-trivial iteration.

5.7 Interpolatory Iteration Formulae

If a function $f(t)$ and all its derivatives $f^{(r)}(t)$ $(r = 1, 2, \ldots, m)$ are known at one oint $t = t_0$, there is is a unique polynomial $P(t)$ of degree at most m, such that

$$P^{(r)}(t_0) = f^{(r)}(t_0), \qquad r = 0, 1, \ldots, m.$$

This *interpolating polynomial*, $P(t)$, can be written as

$$P(t) \equiv f(t_0) + (t - t_0)f'(t_0) + (t - t_0)^2 \frac{f''(t_0)}{2} + \cdots + (t - t_0)^m \frac{f^{(m)}(t_0)}{m!}. \tag{5.20}$$

t follows directly from Taylor's theorem that if f has a continuous $(m + 1)$th lerivative in the neighbourhood of the point $t = t_0$, then the error in this form of nterpolation is

$$f(t) - P(t) = (t - t_0)^{m+1} \frac{f^{(m+1)}(\xi)}{(m + 1)!}, \tag{5.21}$$

vhere $\xi = t_0 + \theta(t - t_0)$ for some $\theta \in (0, 1)$.

There are two possible ways in which interpolation can be used to define iteration formulae, known as direct and inverse interpolation.

(a) Direct interpolation

If the function f is interpolated at a point that is close to a zero of the function then it is reasonable to assume that a zero of the interpolating polynomial will provide a good approximation to that zero. That, in general terms, is the philosoph behind the use of direct interpolation in the context of iteration.

One simple interpolating polynomial is a linear function that interpolates $f(x)$ and $f'(x)$. If the interpolation point is $x = x^{(n)}$, it follows from (5.20) that

$$P(x) \equiv f(x^{(n)}) + (x - x^{(n)})f'(x^{(n)}).$$

If the new approximation $x^{(n+1)}$ is a root of the polynomial equation

$$P(x) = 0,$$

it follows that

$$x^{(n+1)} = x^{(n)} - \frac{f(x^{(n)})}{f'(x^{(n)})}.$$

This is the Newton formula again, and the convergence of Newton's method can now be explained in terms of successive linear interpolations, as illustrated in Figure 10.

Quadratic interpolation, defined by (5.20) with $m = 2$, interpolates $f(x)$, $f'(x)$ and $f''(x)$ at $x = x^{(n)}$; thus it follows that

$$P(x) = f(x^{(n)}) + (x - x^{(n)})f'(x^{(n)}) + (x - x^{(n)})^2 \frac{f''(x^{(n)})}{2}.$$

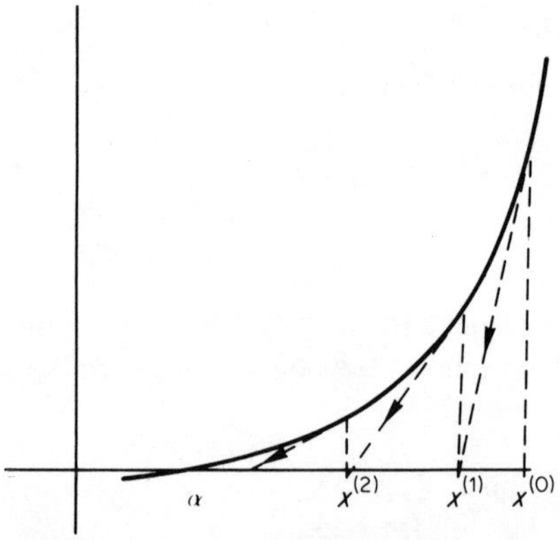

Figure 10 Convergence of Newton's method

Introducing the notation

$$\delta^{(n)} \equiv x^{(n+1)} - x^{(n)}$$

and

$$\gamma^{(n)} \equiv f'(x^{(n)})^2 - 2f(x^{(n)})f''(x^{(n)}), \tag{5.22}$$

it is possible to define the next iterate $x^{(n+1)}$ such that

$$P(x^{(n+1)}) = f(x^{(n)}) + \delta^{(n)}f'(x^{(n)}) + \delta^{(n)^2}\frac{f''(x^{(n)})}{2} = 0;$$

that is

$$\delta^{(n)} = \frac{-f'(x^{(n)}) \pm \gamma^{(n)^{1/2}}}{f''(x^{(n)})}. \tag{5.23}$$

Since it is assumed that $x^{(n)}$ is a good approximation to the true solution, the root of (5.23) that is required is the one with the *smaller absolute value.* This value is achieved when the two terms in the numerator are approximately equal in magnitude, but have opposite signs. It is mentioned in Chapter 1 that an important rule of numerical analysis in general is that such *cancellations of equal and opposite terms lead to a loss of significant figures in the arithmetic.* These situations therefore should be avoided as a matter of principle, and the value computed in an alternative form. If the right-hand side of (5.23) is multiplied by

$$\frac{f'(x^{(n)}) \mp \gamma^{(n)^{1/2}}}{f'(x^{(n)}) \mp \gamma^{(n)^{1/2}}},$$

it follows from (5.22) that

$$\frac{-f'(x^{(n)}) \pm \gamma^{(n)^{1/2}}}{f''(x^{(n)})}\ \frac{f'(x^{(n)}) \mp \gamma^{(n)^{1/2}}}{f'(x^{(n)}) \mp \gamma^{(n)^{1/2}}} = \frac{-2f(x^{(n)})}{f'(x^{(n)}) \mp \gamma^{(n)^{1/2}}}.$$

Thus, if the iteration is computed in the form

$$x^{(n+1)} = x^{(n)} - \frac{2f(x^{(n)})}{f'(x^{(n)}) \mp (f'(x^{(n)})^2 - 2f(x^{(n)})f''(x^{(n)}))^{1/2}}, \tag{5.24}$$

it follows that the smallest root corresponds to an *agreement in sign* in the denominator and so cancellation, with the resultant loss of accuracy, is avoided. One useful feature of the iteration (5.24) and also of Muller's method (introduced in Chapter 6), is that since it involves extracting square roots, it can indicate the presence of *complex conjugate roots* when the iteration is carried out using real arithmetic only. This is not true of Newton's method, which would probably break down in the neighbourhood of complex roots. The convergence of the iteration (5.24) to a real root is illustrated in Figure 11.

Direct interpolation with higher order polynomials is never used as it involves derivatives of third and higher order, and the zeros of even cubic polynomials can

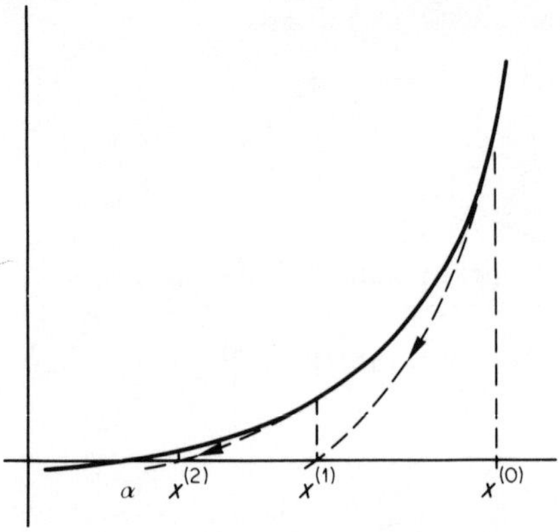

Figure 11 An iteration based on quadratic interpolation

be difficult to obtain – in general it is no easier to find the zeros of a polynomial than to solve an arbitrary non-linear equation.

An alternative to quadratic polynomial interpolation is rational interpolation with functions of the form

$$R(x) = \frac{a - x}{bx + c}\,; \tag{5.25}$$

then if $R(x^{(n+1)}) = 0$, it follows that $x^{(n+1)} = a$. A rational function of the form (5.25), that interpolates $f(x), f'(x)$ and $f''(x)$ at $x = x^{(n)}$, leads to the interaction formula

$$x^{(n+1)} = x^{(n)} - \frac{2f(x^{(n)})f'(x^{(n)})}{2f'(x^{(n)})^2 - f(x^{(n)})f''(x^{(n)})}\,. \tag{5.26}$$

This is known as Halley's method (Traub, 1964), in fact Halley (1694) proposed both (5.16) which he called the *rational form*, and (5.26), which he called the *irrational form*; in spite of this, (5.24) is often attributed to Cauchy (1882). Halley did however use the unstable form (5.23) for his irrational method; round-off error and floating-point arithmetic had no part to play in the pre-computer days of the 17th century, but he did observe that the number of figures correct was tripled at each iteration, which is typical of third-order methods.

(b) Inverse interpolation

If $f(x)$ is a continuous and differentiable function in the neighbourhood of the solution, and if the derivative is non-zero at the solution, then there exists a unique single-valued *inverse function* $F(y)$. An inverse function is defined to be such that if $y = f(x)$ then $x = F(y)$. For example, if

$$f(x) = e^{2x} - 3,$$

then

$$F(y) = \tfrac{1}{2} \ln (y + 3).$$

An example of an inverse function that is not single-valued is shown in Figure 12; it is the inverse of a function in the neighbourhood of a *double root*.

When $F(y)$ is the inverse of $f(x)$, it follows that the equations

$$f(\alpha) = 0$$

and

$$\alpha = F(0),$$

are equivalent. Thus if it is possible to interpolate the inverse function by a polynominal $P(y)$, then an approximate solution can be found by evaluating $P(0)$. The derivatives of the inverse – required both in the formula and in any error bound – can be defined in a straightforward manner from the derivatives of $f(x)$;

$$\frac{\mathrm{d}F}{\mathrm{d}y} = \frac{\mathrm{d}x}{\mathrm{d}y} \equiv \frac{1}{f'}$$

and

$$\frac{\mathrm{d}^2F}{\mathrm{d}y^2} = \frac{\mathrm{d}}{\mathrm{d}y}\left\{\frac{\mathrm{d}x}{\mathrm{d}y}\right\} = \frac{1}{f'}, \qquad \frac{\mathrm{d}}{\mathrm{d}x}\left\{\frac{1}{f'}\right\} = -\frac{f''}{f'^3}.$$

Linear inverse interpolation for $y^{(n)} \equiv f(x^{(n)})$ matches the values of

$$F(y^{(n)}) \equiv x^{(n)} \qquad \text{and} \qquad \frac{\mathrm{d}F(y^{(n)})}{\mathrm{d}y} \equiv -\frac{1}{f'(x^{(n)})}$$

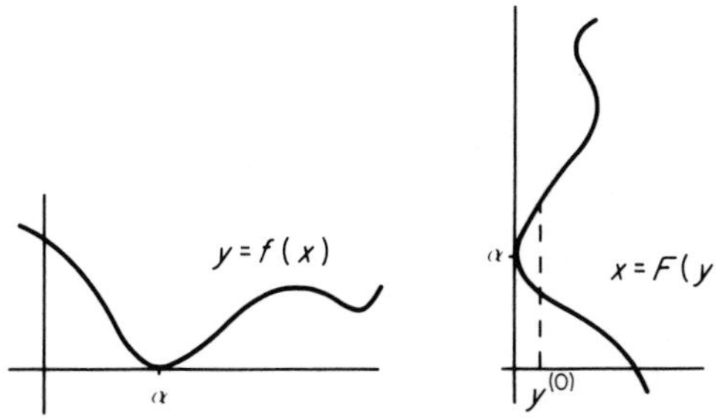

Figure 12 A multiple-valued inverse

is given by

$$P(y) \equiv F(y^{(n)}) + (y - y^{(n)}) \frac{dF(y^{(n)})}{dy}.$$

It then follows that the new approximation $x^{(n+1)} \equiv P(0)$ is given by

$$x^{(n+1)} = x^{(n)} - \frac{f(x^{(n)})}{f'(x^{(n)})},$$

which is Newton's method again. Observe that using this inverse formulation, it follows from (5.21) that

$$\alpha - x^{(n+1)} = \frac{f(x^{(n)})^2}{2} \left(-\frac{f''(\eta)}{f'(\eta)^3} \right),$$

with $\eta = F(\xi)$, where from (5.21) $\xi = (1-\theta) f(x^{(n)})$. Thus as

$$\lim_{n\to\infty} \left[\frac{f(x^{(n)})}{x^{(n)} - \alpha} \right]^2 = f'(\alpha)^2$$

and also η tends to α, it follows that

$$\lim_{n\to\infty} \frac{x^{(n+1)} - \alpha}{(x^{(n)} - \alpha)^2} = \frac{f''(\alpha)}{2f'(\alpha)}, \qquad f'(\alpha) \neq 0.$$

Quadratic inverse interpolation, matching $F(y)$, $dF(y)/dy$ and $d^2F(y)/dy^2$ at $y = y^{(n)} \equiv f(x^{(n)})$, leads to

$$P(y) \equiv F(y^{(n)}) + (y - y^{(n)}) \frac{dF(y^{(n)})}{dy} + (y - y^{(n)})^2 \frac{d^2F(y^{(n)})}{dy^2},$$

which in turn leads to

$$x^{(n+1)} \equiv P(0) = x^{(n)} - \frac{f(x^{(n)})}{f'(x^{(n)})} - \frac{1}{2} \frac{f(x^{(n)})^2 f''(x^{(n)})}{f'(x^{(n)})^3}. \tag{5.27}$$

This formula has been attributed to Chebychev (for example Fröberg, or Rheinboldt, 1974). From (5.21), it follows for this iteration that

$$\alpha - x^{(n+1)} = \frac{f(x^{(n)})^3}{6} \left(\frac{d^3F(\xi)}{dy^3} \right).$$

Since

$$\lim_{n\to\infty} \left[\frac{f(x^{(n)})}{x^{(n)} - \alpha} \right]^3 = f'(\alpha)^3,$$

it follows that

$$\lim_{n\to\infty} \frac{x^{(n+1)}-\alpha}{(x^{(n)}-\alpha)^3} = \frac{f'(\alpha)^3}{6}\left(\frac{\mathrm{d}^3 F(0)}{\mathrm{d}y^3}\right). \tag{5.28}$$

and hence (5.27) is a *third-order method.*

The third-order convergence of the irrational form of *Halley's method* (5.24) can also be verified from the interpolation error. At each stage in the iteration, the interpolating polynomial is

$$P(x) = f(x^{(n)}) + (x - x^{(n)})f'(x^{(n)}) + (x - x^{(n)})^2 \frac{f''(x^{(n)})}{2}, \tag{5.29}$$

for which the interpolation error is

$$f(x) - P(x) = (x - x^{(n)})^3 \frac{f''(\xi)}{6}, \tag{5.30}$$

where ξ lies between x and $x^{(n)}$. If $x = \alpha$ in (5.21), it then follows that

$$-P(\alpha) = (\alpha - x^{(n)})^3 \frac{f'''(\xi)}{6}. \tag{5.31}$$

From the mean value theorem it follows that as $P(x^{(n+1)}) = 0$, then

$$-P(\alpha) = (x^{(n+1)} - \alpha)P'(\eta), \tag{5.32}$$

where η lies between α and $x^{(n+1)}$. By combining (5.31) and (5.32) it follows that

$$\frac{x^{(n+1)}-\alpha}{(x^{(n)}-\alpha)^3} = \frac{1}{6}\frac{f(\xi)}{P'(\eta)}.$$

As the iteration converges, it follows that both η and ξ tend to α; in addition (5.29) leads to

$$\lim_{n\to\infty} P'(\eta) = \lim_{n\to\infty} \frac{\mathrm{d}}{\mathrm{d}x} P(x^{(n)}) = \lim_{n\to\infty} f'(x^{(n)}) = f'(\alpha).$$

Thus

$$\lim_{n\to\infty} \frac{x^{(n+1)}-\alpha}{(x^{(n)}-\alpha)^3} = -\frac{1}{6}\frac{f'''(\alpha)}{f'(\alpha)}, \tag{5.33}$$

and hence the convergence is third order. This method – together with the rational form (5.26) – has been extensively studied by Traub (1964).

Exercises

5.28. Verify that if

$$R(x) = \frac{a - x}{bx + c}$$

interpolates $f(x), f'(x)$ and $f''(x)$ at $x = x^{(n)}$, then $x^{(n+1)}$ such that $R(x^{(n+1)}) = 0$ is given by (5.26).

5.29. Prove that for $j > 1$,

$$\frac{d^j F}{dy^j} = \frac{1}{f'(x)} \left\{ \frac{d}{dx} \right\}^{j-1} \frac{1}{f'(x)}. \tag{5.34}$$

5.30. Evaluate the asymptotic error constant of (5.27) in terms of $f'(\alpha), f''(\alpha)$ and $f'''(\alpha)$ using (5.27) and (5.28).

5.31. Verify that (5.26) is obtained from local interpolation in terms of hyperbolae whereas (5.27) is from local interpolation by parabolae.

5.8 Newton's Method for Systems of Equations

In the preceding section, Newton's method for solving a single equation is developed in terms of local linear interpolation. It is possible to construct Newton's method for systems of equations in an analogous manner by interpolating each component of $\mathbf{f}(\mathbf{x})$ by a linear polynomial of $x_1, \ldots, x_N$. If the polynomial matches the function value and *all* first derivatives at the point $\mathbf{x} = \mathbf{x}^{(n)}$, it follows that

$$\mathbf{P}(\mathbf{x}) \equiv [P_1(\mathbf{x}), \ldots, P_N(\mathbf{x})]^{\mathrm{T}}$$

can be written as

$$P_i(\mathbf{x}) = f_i^{(n)} + \sum_{j=1}^{N} (x_j - x_j^{(n)}) f_{ij}^{(n)}, \qquad i = 1, \ldots, N. \tag{5.35}$$

The notation defined by

$$f_i^{(n)} \equiv f_i(x^{(n)}) \qquad \text{with} \qquad \mathbf{f}^{(n)} \equiv [f_1^{(n)}, \ldots, f_N^{(n)}]^{\mathrm{T}}$$

and

$$f_{ij}^{(n)} \equiv \frac{\partial f_i(x^{(n)})}{\partial x_j} \qquad \text{with} \qquad J^{(n)} \equiv \{f_{ij}^{(n)}\}$$

is used in both this section and the next to simplify some of the formulae. It follows from Taylor's theorem that the error in this form of linear interpolation is given by

$$f_i(\mathbf{x}) - P_i(\mathbf{x}) = \tfrac{1}{2}(\mathbf{x} - \mathbf{x}^{(n)})^{\mathrm{T}} H_i(\boldsymbol{\xi}_i)(\mathbf{x} - \mathbf{x}^{(n)}), \tag{5.36}$$

where $H_i(\boldsymbol{\xi}_i)$ is the matrix of second derivatives – the *Hessian matrix* – of the function $f_i(\mathbf{x})$, evaluated at some point

$$\boldsymbol{\xi}_i = \theta_i \mathbf{x} + (1 - \theta_i)\mathbf{x}^{(n)}, \qquad \theta_i \in (0, 1).$$

The systems of equations (5.35) can be written as

$$\mathbf{P}(\mathbf{x}) = \mathbf{f}^{(n)} + J^{(n)}(\mathbf{x} - \mathbf{x}^{(n)}). \tag{5.37}$$

At the nth stage of the iteration, it is necessary to solve the linear system $\mathbf{P}(\mathbf{x}) = 0$ in order to obtain the next iterate $\mathbf{x}^{(n+1)}$. Thus it follows from (5.37) that it is necessary to solve the linear system

$$J^{(n)}\boldsymbol{\delta}^{(n)} = -\mathbf{f}^{(n)} \tag{5.38a}$$

in order to determine the correction

$$\boldsymbol{\delta}^{(n)} \equiv \mathbf{x}^{(n+1)} - \mathbf{x}^{(n)}. \tag{5.38b}$$

Note that the linear system (5.38a) should be solved by one of the methods outlined in Chapter 2; under no circumstances should the correction be computed by explicitly inverting the matrix $J^{(n)}$.

It is possible to show that the method is quadratically convergent, like its single-equation counterpart. The iteration can be written as

$$\mathbf{x}^{(n+1)} = \boldsymbol{\varphi}(\mathbf{x}^{(n)}),$$

where

$$\boldsymbol{\varphi}(\mathbf{x}) = \mathbf{x} - J(\mathbf{x})^{-1}\,\mathbf{f}(\mathbf{x}).$$

The columns of the Jacobian matrix of $\varphi(\mathbf{x})$ can be written as

$$\mathbf{e}_i - J(\mathbf{x})^{-1}\frac{\partial}{\partial x_i}[\mathbf{f}(\mathbf{x})] - G_i(\mathbf{x})\mathbf{f}(\mathbf{x}), \qquad i = 1, \ldots, N,$$

where $\mathbf{e}_i$ is the ith unit vector,

$$\frac{\partial}{\partial x_i}[\mathbf{f}(\mathbf{x})]$$

is the vector of ith derivatives of the components of $\mathbf{f}(\mathbf{x})$ and $G_i(\mathbf{x})$ is a matrix involving second derivatives of $\mathbf{f}(\mathbf{x})$. Then (see Exercise 5.28) $J_\varphi(\boldsymbol{\alpha}) = 0$, and by analogy with the analysis of Section 5.6, the method is at least second order. A more rigorous result that in general

$$\| \mathbf{x}^{(n+1)} - \boldsymbol{\alpha} \| = O(\| \mathbf{x}^{(n)} - \boldsymbol{\alpha} \|^2),$$

is beyond the scope of this book and can be found, for example, in Ortega (1972).

Numerical example (Newton's method for a 3 x 3 system of equations). The system of equations

$$f_1(\mathbf{x}) \equiv -\tfrac{1}{2}x_1 + \tfrac{1}{3}(x_2^3 + x_3^3) + 0.3 = 0,$$
$$f_2(\mathbf{x}) \equiv \tfrac{1}{2}x_1x_2 - \tfrac{1}{2}x_2 + \tfrac{1}{5}x_3^5 = 0,$$
$$f_3(\mathbf{x}) \equiv \tfrac{1}{3}x_2^3 - \tfrac{1}{4}x_3 + 0.04 = 0$$

has the Jacobian matrix

$$J(\mathbf{x}) = \begin{bmatrix} -\frac{1}{2} & x_2^2 & x_3^2 \\ \frac{1}{2}x_2 & \frac{1}{2}(x_1 - 1) & x_3^4 \\ 0 & x_2^2 & -\frac{1}{4} \end{bmatrix}.$$

The sequence $\{\mathbf{x}^{(n)}\}$ defined by

$$\mathbf{x}^{(n+1)} = \mathbf{x}^{(n)} - J(\mathbf{x}^{(n)})^{-1} f(\mathbf{x}^{(n)}), \qquad n = 0, 1, \ldots$$

with $\mathbf{x}^{(0)} = [0, 0, 0]^{\mathrm{T}}$ and computed as in (5.38a) and (5.38b), is given in Table 4. As a comparison the iteration was repeated with $J(\mathbf{x}^{(n)})$ replaced by the constant matrix

$$J(\mathbf{x}^{(0)}) = \begin{bmatrix} -\frac{1}{2} & & \\ & -\frac{1}{2} & \\ & & -\frac{1}{4} \end{bmatrix},$$

the results of this iteration are given in Table 5. A more accurate approximation to the Jacobian matrix at the solution is given by $J(\mathbf{x}^*)$, where $\mathbf{x}^* = [0.6, 0, 0]^{\mathrm{T}}$, that is

$$J(\mathbf{x}^*) = \begin{bmatrix} -\frac{1}{2} & & \\ & -\frac{1}{5} & \\ & & -\frac{1}{4} \end{bmatrix}. \tag{5.39}$$

The results of this alternative iteration are the same as for Newton's method.

Table 4 Newton's method

n	x_1	x_2	x_3
1	0.6000000	0.	0.1600000
2	0.6027307	0.0001049	0.1600000
3	0.6027307	0.0001056	0.1600000
4	0.6027307	0.0001056	0.1600000

A standard modification of the basic Newton method is illustrated by the above example; that is to compute the sequence

$$\mathbf{x}^{(n+1)} = \mathbf{x}^{(n)} - J(\mathbf{X})^{-1} \mathbf{f}(\mathbf{x}^{(n)}), \qquad n = 0, 1, \ldots$$

with a fixed value for $\mathbf{X}$, thus saving in the arithmetic at each iteration as it is not necessary to re-factorize the matrix $J(\mathbf{X})$. As the numerical example shows, if $J(\mathbf{X})$ is significantly different from $J(\mathbf{x}^{(n)})$, this method can be slow and ultimately wasteful in computer time. A sensible strategy is therefore to recompute the

able 5 Iteration with a constant Jacobian matrix

	x_1	x_2	x_3
1	0.6000000	0.	0.1600000
2	0.6027307	0.0000419	0.1600000
3	0.6027307	0.0000672	0.1600000
4	0.6027307	0.0000825	0.1600000
5	0.6027307	0.0000916	0.1600000
6	0.6027307	0.0000972	0.1600000
7	0.6027307	0.0001005	0.1600000
8	0.6027307	0.0001025	0.1600000
9	0.6027307	0.0001037	0.1600000
0	0.6027307	0.0001045	0.1600000
1	0.6027307	0.0001049	0.1600000
2	0.6027307	0.0001052	0.1600000
3	0.6027307	0.0001053	0.1600000
4	0.6027307	0.0001054	0.1600000
5	0.6027307	0.0001055	0.1600000
6	0.6027307	0.0001055	0.1600000

cobian if the convergence appears slow. Clearly this modification of Newton's ethod has to be used with caution. It has proved extremely useful in applications here similar sets of non-linear equations are solved sequentially and the solution f one set provides a *good* initial approximation for the next set. As illustrated in ection 1.3, this situation arises frequently in the numerical solution of non-linear ifferential equations (Gear, 1971), where a system of equations of the form

$$\mathbf{f}(\mathbf{x}, t, \mathbf{X}) = 0$$

solved for $\mathbf{x}$ as the parameter t (time) is incremented. At any value of t; $\mathbf{x} \equiv \mathbf{x}(t)$ the desired solution, and $\mathbf{X} \equiv \mathbf{x}(t - \Delta t)$ the computed solution at the previous me step. Thus at each step, provided t is sufficiently small, $\mathbf{x}^{(0)} = \mathbf{X}$ is a good first pproximation. In many implementations it has been found possible to retain the me Jacobian matrix for several time steps, recomputing it only when the iteration ils to converge sufficiently rapidly.

The rôle of non-linear algebraic equations in the numerical solution of differential quations is mentioned again during the discussion of *quasi-Newton methods.*

Exercises

.32. Verify that

$$J(\mathbf{x})^{-1} \frac{\partial}{\partial x_i} [\mathbf{f}(\mathbf{x})] = \mathbf{e}_i, \qquad i = 1, \ldots, N.$$

.33. Prove that if, for each component of $\mathbf{f}$, the corresponding matrix $H_i(\boldsymbol{\xi})$ is ositive definite and uniformly bounded for all $\boldsymbol{\xi}$, and if $\nabla f_i(\boldsymbol{\eta})$ is non-vanishing or all $\boldsymbol{\eta}$, then it follows from (5.36) and the mean value theorem that there exists a positive constant C such that

$$\| \mathbf{x}^{(n+1)} - \boldsymbol{\alpha} \|_2 \leqslant C \| \mathbf{x}^{(n)} - \boldsymbol{\alpha} \|_2^2.$$

5.34. Prove that

$$G_i(\mathbf{x}) \equiv -J(\mathbf{x})^{-1} \frac{\partial}{\partial x_i} [J(\mathbf{x})] J(\mathbf{x})^{-1}, \qquad i = 1, \ldots, N.$$

5.35. One simplification of the basic Newton method is the so-called *non-linear Jacobi method*; the full Jacobian matrix $J(\mathbf{x})$ is replaced by the diagonal terms only. Thus

$$\varphi_i(\mathbf{x}) = x_i - \frac{f_i(\mathbf{x})}{(\partial f_i(\mathbf{x})/\partial x_i)}, \qquad i = 1, \ldots, N.$$

Prove, using the contraction mapping theorem, that this method converges for any starting approximation if:

(a) $J(\mathbf{x})$ is uniformily and strictly diagonally dominant for all $\mathbf{x}$; that is, there exist $\rho < 1$ such that

$$\sum_{j \neq i} \left| \frac{\partial f_i}{\partial x_j} \right| \leqslant \rho \left| \frac{\partial f_i}{\partial x_i} \right| ;$$

(b) the second derivatives satisfy

$$\frac{|f_i|}{(\partial f_i/\partial x_i)^2} \sum_{j=1}^{N} \left| \frac{\partial^2 f_i}{\partial x_i \partial x_j} \right| + \rho \leqslant L$$

for some $L < 1$.

5.36. Explain the results in Tables 4 and 5, in particular verify that the constant Jacobian matrix (5.39) and Newton's method lead to identical iterations, not just the same to 7 significant figures.

5.9 Brown's Method

A useful modification of Newton's method into a Gauss–Seidel type of method has been suggested by Brown (1967). Newton's method relies on the simultaneous local interpolation of all the components of $\mathbf{f}(\mathbf{x})$ by a linear system $\mathbf{P}(\mathbf{x})$. The linear equations $\mathbf{P}(\mathbf{x}) = 0$ are then solved in order to find the next iterate. The modificatio suggested by Brown is linearize the components sequentially, using each linear equation to eliminate a single component of the solution from the remaining non-linear equations as in Gauss elimination. The system eventually reduces to a single non-linear equation in a single unknown, to which one step of the Newton iteration is then applied. The new values of all the eliminated components are then obtained in the reverse order by back substitution.

The success of the method may well depend on the ordering of the components of $\mathbf{f}$ and $\mathbf{x}$ in any particular problem. Assuming that the components of both $\mathbf{f}$ and are considered in the reverse order, the algorithm at the nth stage can be written as

(1) Interpolate $f_N(\mathbf{x})$ by $P_N(\mathbf{x})$ and solve $P_N(\mathbf{x}) = 0$ for x_N in terms of

$x_1, \ldots, x_{N-1}$; thus

$$P_N(\mathbf{x}) \equiv f_N^{(n)} + \sum_{j=1}^{N} (x_j - x_j^{(n)}) f_{Nj}^{(n)},$$

hence

$$x_N = x_N^{(n)} - (f_{NN}^{(n)})^{-1} \left\{ \sum_{j=1}^{N-1} (x_j - x_j^{(n)}) f_{Nj}^{(n)} + f_N^{(n)} \right\}$$

$$\equiv X_N(x_1, \ldots, x_{N-1}).$$

(2) For $i = N - 1, \ldots, 2$, define

$$g_i(x_1, \ldots, x_i) \equiv f_i(x_1, \ldots, x_i, X_{i+1}, \ldots, X_N) \tag{5.40}$$

and interpolate g_i by $P_i(x_1, \ldots, x_i)$ and solve $P_i = 0$ for x_i in terms of $x_1, \ldots, x_{i-1}$. Thus

$$P_i(x_1, \ldots, x_i) \equiv g_i^{(n)} + \sum_{j=1}^{i} (x_j - x_j^{(n)}) g_{ij}^{(n)}$$

where $g_{ij}^{(n)}$ and $g_i^{(n)}$ are analogous to $f_{ij}^{(n)}$ and $f_i^{(n)}$ respectively. Then

$$x_i = x_i^{(n)} - (g_{ii}^{(n)})^{-1} \left\{ \sum_{j=1}^{i-1} (x_j - x_j^{(n)}) g_{ij}^{(n)} + g_i^{(n)} \right\}$$

$$\equiv X_i(x_1, \ldots, x_{i-1}). \tag{5.41}$$

It follows from (5.40) that the derivatives

$$g_{ij}^{(n)}, \qquad j = 1, \ldots, i$$

are evaluated from the previously computed values

$$g_{kj}^{(n)} \qquad k = i + 1, \ldots, N; \quad j = 1, \ldots, k$$

using the relationships

$$\frac{\partial g_i}{\partial x_j} = \frac{\partial f_i}{\partial x_j} + \sum_{k=i+1}^{N} \frac{\partial f_i}{\partial x_k} \frac{\partial X_k}{\partial x_j}$$

$$= \frac{\partial f_i}{\partial x_j} + \sum_{k=i+1}^{N} \frac{\partial f_i}{\partial x_k} \left(\frac{g_{kj}^{(n)}}{g_{kk}^{(n)}} \right), \qquad j = 1, \ldots, i; \quad g_{Nj}^{(n)} \equiv f_{Nj}^{(n)}. \tag{5.42}$$

(3) Once the functions X_i ($i = N, \ldots, 2$) have been computed it only remains to perform a single Newton iteration on the function

$$g_1(x_1) \equiv f_1(x_1, X_2, \ldots, X_N);$$

thus

$$x_1^{(n+1)} = x_1^{(n)} - \frac{g_1^{(n)}}{g_{11}^{(n)}},$$

where $g_{11}^{(n)}$ is evaluated using (5.42).

(4) The values of $x_i^{(n+1)}$ $(i + 2, \ldots, N)$ are determined by substitution from (5.41) as

$$x_i^{(n+1)} = X_i(x_1^{(n+1)}, \ldots, x_{i-1}^{(n+1)}) \qquad i = 2, \ldots, N.$$

Computing g_{ij} by differences uses fewer function evaluations per step than Newton's method – $\frac{1}{2}(N^2 + 3N)$ compared to $N^2 + N$ – and it retains the (ultimate) quadratic convergence of Newton's method.

Numerical example (The solution of the 3 x 3 system given in (5.8b)). In this example

$$P_3(\mathbf{x}) = f_3^{(n)} + (x_1 - x_1^{(n)})(x_3^{(n)} - 2) + (x_3 - x_3^{(n)})x_1^{(n)},$$

$$X_3(x_1, x_2) + x_3^{(n)} - \frac{1}{x_1^{(n)}}\left[(x_3^{(n)} - 2)x_1 + k\right],$$

$$P_2(x_1, x_2) = f_2^{(n)} + (x_1 - x_1^{(n)}) + 5(x_2 - x_2^{(n)}),$$

and

$$X_2(x_1) = -\tfrac{1}{5}(1 + x_1).$$

This example illustrates the significance of the method in solving systems containing several linear equations. Finally

$$g_1(x_1) = x_1^2 + \tfrac{1}{25}(1 + x_1)^2 - \left\{x_3^{(n)} - \frac{1}{x_1^{(n)}}\left[(x_3^{(n)} - 2)x_1 + k\right]\right\} - 2.$$

The results of this iteration with $k = \pm 1$ with an initial approximation $x^{(0)} = [-2, 0, 2]^{\mathrm{T}}$ are given in Table 6.

Table 6 Brown's method

k	n	x_1	x_2	x_3
−1	0	−2.0000000	0.00000000	+2.0000000
	1	−1.8676471	0.17352941	+1.5000000
	2	−1.8524785	0.17049570	+1.4605061
	3	−1.8523169	0.17046338	+1.4601356
	4	−1.8523169	0.17046338	+1.4601356
1	0	−2.0000000	0.00000000	+2.0000000
	1	−2.1127451	0.22254902	+2.5000000
	2	−2.1039777	0.22079553	+2.4753928
	3	−2.1039373	0.22078746	+2.4752993
	4	−2.1039373	0.22078746	+2.4752993

5.10 Global Convergence of Newton's Method

The orders of convergence derived in the preceding sections are *local* results in e sense that it is assumed that $|x^{(n)} - \alpha|$ is small and that

$$\lim_{n\to\infty} (x^{(n)} - \alpha) = 0.$$

he contraction mapping, on the other hand, provided a *global* convergence eorem, since if the iteration function is a contraction mapping, convergence is uaranteed irrespective of the precise value of the starting approximation. Thus in rder to prove the global convergence of Newton's method it is necessary to find e conditions under which the function

$$\varphi(x) = x - \frac{f(x)}{f'(x)}$$

a contraction mapping. It follows from the mean value theorem that one sufficient ondition is

$$\left|\frac{f(x)f''(x)}{f'(x)^2}\right| \leqslant L < 1.$$

n alternative result that has a straightforward extension to systems and that does ot depend on a knowledge of the second derivative is given below:

heorem *If the solution* α *of* $f(x) = 0$ *lies in an interval* I *in which*

$$f'(x) \neq 0 \qquad \text{and continuous for all } x \in I \tag{5.43}$$

nd

$$f(a) - f(b) \geqslant f'(b)(a - b) \qquad \text{for all } a, b \in I, \tag{5.44}$$

hen, for any $x^{(0)} \in I$, *the Newton iteration leads to a sequence* $\{x^{(n)}\}$ *which atisfies exactly one of the following conditions;*

(1) $x^{(n)} \in I$ *and* $\{x^{(n)}\}$ *tends monotonically to* α;
(2) $f(x^{(0)})f(x^{(1)}) < 0$ *and* $x^{(1)}, x^{(2)}, \ldots$ *converges monotonically to* α; *or*
(3) $x^{(1)} \notin I$. △

Note that if $I \equiv \mathbb{R}$, then the third alternative does not exist.

Corollary 1 *The* convexity condition (5.44) *can be replaced by the* concavity condition

$$f(a) - f(b) \leqslant f'(b)(a - b) \qquad \text{for all } a, b \in I \tag{5.45}$$

and the theorem is still valid. △

Examples of convex and concave functions are shown in Figure 13.

Corollary 2 *The conditions (5.44) or (5.45), when taken with (5.43), can be*

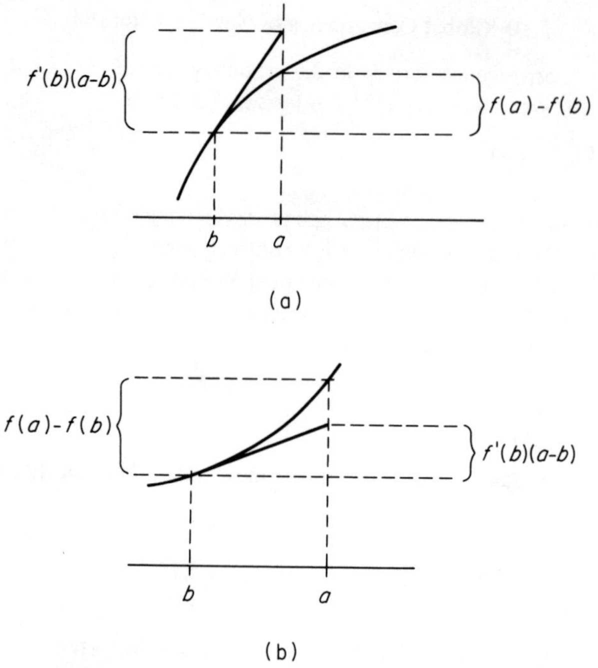

Figure 13 (a) A concave function; (b) a convex function

replaced by the condition

$$f''(x) \neq 0 \qquad \textit{and is continuous for all } x \in I. \quad \Delta \tag{5.46}$$

Proof of Theorem Assume without loss of generality that $f'(x) > 0$ for all $x \in I$. The proof for $f'(x) < 0$ is left as Exercise 5.38. Since

$$x^{(n+1)} = x^{(n)} - \{f'(x^{(n)})\}^{-1} f(x^{(n)}), \qquad n = 1, 2, \ldots,$$

it follows from (5.44) that if $x^{(n)}, x^{(n+1)} \in I$, then

$$-f(x^{(n)}) = f'(x^{(n)})(x^{(n+1)} - x^{(n)}) \leqslant f(x^{(n+1)}) - f(x^{(n)}) \tag{5.47}$$

and hence

$$f(x^{(n+1)}) \geqslant 0, \qquad n = 1, 2, \ldots.$$

Similarly

$$-f(x^{(n+1)}) \geqslant f'(x^{(n+1)})(\alpha - x^{(n+1)}),$$

and so

$$x^{(n+1)} - \alpha \geqslant \frac{f(x^{(n+1)})}{f'(x^{(n+1)})} \geqslant 0, \qquad n = 0, 1, \ldots. \tag{5.48}$$

It also follows that

$$x^{(n+2)} = x^{(n+1)} - \{f'(x^{(n+1)})\}^{-1} f(x^{(n+1)}) \leqslant x^{(n+1)}, \qquad n = 0, 1, \ldots, \tag{5.49}$$

and so

$$\alpha \leqslant x^{(n+2)} \leqslant x^{(n+1)}, \qquad n = 0, 1, \ldots .$$

The sequence $x^{(1)}, x^{(2)}, \ldots$ is therefore bounded and monotone; hence a limit point exists and as

$$\lim_{n\to\infty} x^{(n)} = \lim_{n\to\infty} \left\{ x^{(n)} - \frac{f(x^{(n)})}{f'(x^{(n)})} \right\},$$

the iteration converges to a solution of $f(x) = 0$. The condition $f'(x) \neq 0$ assures the uniqueness.

It follows from (5.49) that if $x^{(1)} \in I$, then convergence of the sequence $x^{(1)}$, $x^{(2)}, \ldots$ is monotone and the three possibilities follow directly:

(1) If $x^{(0)} > \alpha$, then from (5.48), $f(x^{(0)}) > 0$ and the sequence $x^{(0)}, x^{(1)}$, $x^{(2)}, \ldots$ converges monotonically.
If $x^{(0)} < \alpha$, then either
(2) $x^{(1)} \in I$ and convergence of $x^{(1)}, x^{(2)}, \ldots$ is monotone; or
(3) $x^{(1)} \notin I$. Δ

The three alternatives are illustrated in Figure 14.

The theorem can be formulated for systems of equations: then (5.43) is replaced by

$$[J(\mathbf{x})]^{-1} \geqslant 0 \qquad \text{for all } \mathbf{x} \in \mathbb{R}^N$$

and (5.44) by

$$\mathbf{f}(\mathbf{a}) - \mathbf{f}(\mathbf{b}) \geqslant J(\mathbf{b})(\mathbf{a} - \mathbf{b}) \qquad \text{for all } \mathbf{a}, \mathbf{b} \in \mathbb{R}^N;$$

the inequalities are defined component by component.

As the interval I has been replaced by the whole of $\mathbb{R}^N$, it follows that the third alternative is deleted. The only modifications necessary in the proofs are notational. Probably the most important convergence result for systems of equations is, however, the Kantorovich theorem (see, for example, Goldstein, 1967), but a discussion is beyond the scope of this book.

The conditions (5.43) and (5.46) can be combined to give

$$f'(x)f''(x) \neq 0 \qquad \text{for all } x \in I; \tag{5.50}$$

this is known as the *Fourier condition* (Ostrowski, 1966). Traub (1964) has generalized the Fourier condition to determine the conditions under which convergence of higher-order iterations is monotone. The monotonicity follows from the local error estimate; for example, it follows from (5.33) that a *necessary*

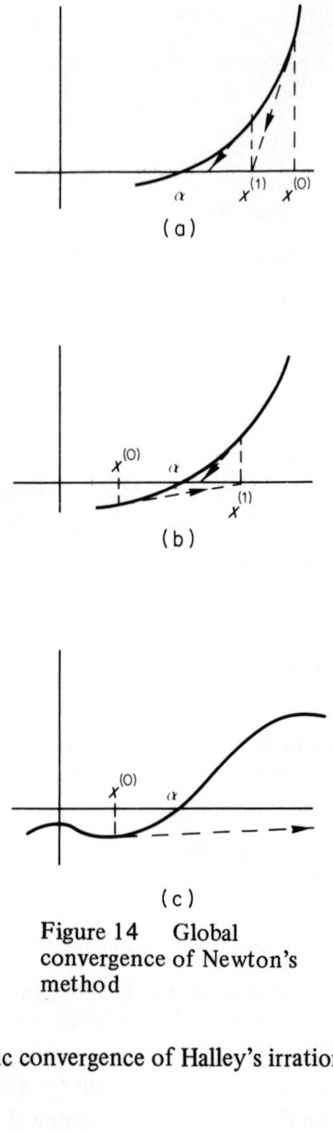

Figure 14 Global convergence of Newton's method

condition for the monotonic convergence of Halley's irrational form is that

$$f'''(x)f'(x) < 0.$$

Fourier conditions have also been derived for Halley's rational method (Davies and Dawson, 1975).

When $f(x)$ is convex (or concave) and the convergence of Newton's method is monotonic, it is possible to construct an additional approximating sequence $\{y_n\}$ such that

$$y^{(n+1)} = y^{(n)} - \frac{f(y^{(n)})}{f'(x^{(n)})}, \qquad n = 0, 1, \ldots, \tag{5.51}$$

where $\{x^{(n)}\}$ is the sequence generated by the Newton iteration. Then $x^{(n)}$ and $y^{(n)}$ provide upper and lower bounds on the solution. This device, due to Fourier (see Historical Appendix), is an example of a *bracketing method.* Such methods are considered in more detail in Section 6.4.

Exercises

5.37. Prove that (5.43) and (5.46) imply either (5.44) or (5.45).

5.38. Verify that if $f(x)$ is a *convex* function such that $f'(x) < 0$, then any monotone convergence is from below, i.e. $x^{(n)} \leqslant x^{(n+1)} \leqslant \alpha$.

5.39. Determine the direction of the monotone convergence for a *concave* function.

5.40. Assume that $f'(x)$ and $f''(x)$ are positive in some interval I containing a root. Verify that $f(x)$ and $f'(x)$ are monotonically increasing functions in I; that is, $f(x)$ is convex. Hence, using the mean value theorem, prove that for any $x^{(0)} > \alpha$ and $y^{(0)} < \alpha$ such that $x^{(0)}, y^{(0)} \in I$, the sequence of Newton iterates $\{x^{(n)}\}$ is monotonically decreasing and converges to α, whereas the sequence $\{y^{(n)}\}$ defined by (5.51) is monotonically increasing and converges to α.

6

Multi-step Iteration Formulae

6.1 Introduction

Multistep iteration formulae (alternatively known as *multi-point formulae* or *formulae with memory*) re-use function values computed at earlier stages of the iteration. The object in re-using old information is to obtain a more efficient method in the sense that it is usually cheaper and faster to access a stored value of the function $f(x)$ rather than compute (say) a value of $f'(x)$.

For example, if it is assumed that

$$f'(x^{(n)}) \approx \frac{f(x^{(n)}) - f(x^{(n-1)})}{x^{(n)} - x^{(n-1)}}, \tag{6.1}$$

Newton's method can be replaced by the formula

$$x^{(n+1)} = x^{(n)} - f(x^{(n)})\left\{\frac{x^{(n)} - x^{(n-1)}}{f(x^{(n)}) - f(x^{(n-1)})}\right\}, \qquad n = 0, 1, \ldots .$$

The use of this formula is known as the *secant method*, and it clearly requires only one function evaluation per iteration step. The secant method is clearly an iteration of the form

$$x^{(n+1)} = \varphi(x^{(n)}, x^{(n-1)}), \qquad n = 0, 1, \ldots,$$

where φ is a single-valued function of two variables. That is, φ is a mapping from $\mathbb{R}^2$ into $\mathbb{R}$ and so as the domain and range are different, the contraction mapping theorem is not valid. In addition, the order of convergence can no longer be determined according to the rules set out in Section 5.6. Therefore neither the local nor global convergence theory of the previous chapter is particularly appropriate.

One important class of multi-step iteration formulae is derived by polynomial interpolation in much the same way as the methods of Newton and Halley are derived. It is first necessary to define the form of interpolating polynomials used.

For any function $f(t)$ it is possible to define *divided differences* with respect to a set of distinct points t_i $(i = 1, \ldots, m)$. Let

$$f_i \equiv f(t_i), \qquad i = 1, \ldots, m;$$

then

$$f_{i,i-1} \equiv \frac{f_i - f_{i-1}}{t_i - t_{i-1}}, \qquad i = 2, \ldots, m$$

and in general

$$f_{i,\ldots,i-j} \equiv \frac{f_{i,\ldots,i-j+1} - f_{i-1,\ldots,i-j}}{t_i - t_{i-j}}, \qquad \begin{cases} i = 3, \ldots, m \\ j = 2, \ldots, i-1. \end{cases}$$

There is a unique polynomial, of degree at most $m - 1$, that interpolates the function f at the m points t_i $(i = 1, \ldots, m)$. The interpolating polynomial can be written as

$$f(t) \equiv f_m + (t - t_m)f_{m,m-1} + (t - t_m)(t - t_{m-1})f_{m,m-1,m-2} + \cdots + \prod_{i=2}^{m} (t - t_i) f_{m,m-1,\ldots,1}. \tag{6.2}$$

This is known as the *divided difference form* for the interpolating polynomial; it is not unique and other forms, such as the *Lagrange form*, can be found in most general texts on numerical analysis.

If $f(t)$ is m times continuously differentiable, the error in such interpolation can be written as

$$f(t) - P(t) = \prod_{i=1}^{m} (t - t_i) \frac{1}{m!} \frac{\mathrm{d}^m f(\xi)}{\mathrm{d}t^m}, \tag{6.3}$$

where

$$\xi \equiv \xi(t) \in \mathrm{Int}\,\{t, t_1, \ldots, t_m\}.$$

The notation $\mathrm{Int}\{x\}$ means the interval (a, b) where $a = \min\{x\}$ and $b = \max\{x\}$. Interpolating polynomials can be defined when some of the points are coincident, that is when function values and derivatives are interpolated.

The difference approximation (6.1) is by no means the only one possible; a more accurate replacement is, in divided difference notation,

$$f_{n,n-1} + f_{n,n-2} - f_{n-1,n-2}.$$

This leads to the iteration formula

$$x^{(n+1)} = x^{(n)} - \frac{f^{(n)}}{f_{n,n-1} + f_{n,n-2} - f_{n-1,n-2}}, \qquad n = 0, 1, \ldots, \tag{6.4}$$

where $f^{(n)} \equiv f(x^{(n)})$ and $f_{n,n-1}$, etc. are divided differences of $f^{(n)}, f^{(n-1)}$, etc.

This is an iteration in the form

$$x^{(n+1)} = \varphi(x^{(n)}, x^{(n-1)}, x^{(n-2)}).$$

In the divided difference notation the secant method can be written as

$$x^{(n+1)} = x^{(n)} - \frac{f^{(n)}}{f_{n,n-1}}, \qquad n = 0, 1, \ldots .$$

Exercises

6.1. Verify, using a divided difference approximation to the derivative of the inverse function F, that the iteration formula

$$x^{(n+1)} = x^{(n)} - f^{(n)}\left\{\frac{1}{f_{n,n-1}} + \frac{1}{f_{n,n-2}} - \frac{1}{f_{n-1,n-2}}\right\} \tag{6.5}$$

is a difference approximation to Newton's method analogous to (6.4).

6.2. Verify by means of Taylor's theorem that the difference approximations in (6.4) and (6.5) are of similar accuracy.

6.2 Interpolatory Iteration Formulae

As in Chapter 5, two forms of interpolation, known as direct and inverse interpolation, can be used to define iteration formulae.

(a) Direct Interpolation

The simplest form of interpolation employs a linear polynomial. If the interpolation points are $x = x^{(n)}$ and $x = x^{(n-1)}$, it follows from (6.2) that

$$P(x) = f^{(n)} + (x - x^{(n)})f_{n,n-1}.$$

If the new approximation $x^{(n+1)}$ is a root of

$$P(x) = 0,$$

it follows that

$$x^{(n+1)} = x_n - \frac{f^{(n)}}{f_{n,n-1}}. \tag{6.6}$$

This is the secant method again; the iteration is illustrated in Figure 15.

Quadratic interpolation requires three data points; if these are taken as $x^{(n)}$, $x^{(n-1)}$ and $x^{(n-2)}$, then $x^{(n+1)}$, the new approximation to the root, is given by

$$f^{(n)} + (x^{(n+1)} - x^{(n)})f_{n,n-1} + (x^{(n+1)} - x^{(n)})(x^{(n+1)} - x^{(n-1)})f_{n,n-1,n-2} = 0. \tag{6.7}$$

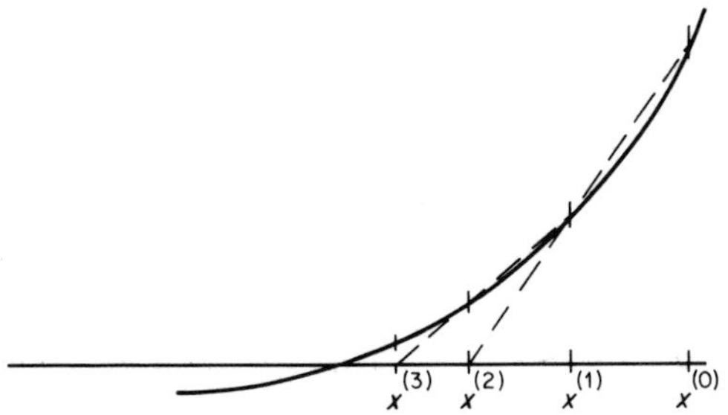

Figure 15 Convergence of the secant method

Introducing the notation

$$\delta^{(n)} \equiv x^{(n+1)} - x^{(n)}$$

and

$$\gamma_n \equiv f_{n,n-1} + (x^{(n)} - x^{(n-1)})f_{n,n-1,n-2},$$

equation (6.7) can be written as

$$f^{(n)} + \delta^{(n)}f_{n,n-1} + \delta^{(n)}(\delta^{(n)} + (x^{(n)} - x^{(n-1)}))f_{n,n-1,n-2} = 0.$$

This has the solutions

$$\delta^{(n)} = \frac{-\gamma_n \pm (\gamma_n^2 - 4f^{(n)}f_{n,n-1,n-2})^{1/2}}{2f_{n,n-1}}. \tag{6.8}$$

In a manner analogous to that of Section 5.7, the right-hand side of (6.8) is multiplied by

$$\frac{\gamma_n \mp (\gamma_n^2 - 4f^{(n)}f_{n,n-1,n-2})^{1/2}}{\gamma_n \mp (\gamma_n^2 - 4f^{(n)}f_{n,n-1,n-2})^{1/2}}$$

and the iteration formula becomes

$$x^{(n+1)} = x^{(n)} - \frac{2f^{(n)}}{\gamma_n \mp (\gamma_n^2 - 4f^{(n)}f_{n,n-1,n-2})^{1/2}}. \tag{6.9}$$

Now, the smallest root corresponds to *agreement of sign* in the denominator, and so there is no cancellation with its resulting loss of accuracy. This method is known as *Muller's method* (Muller, 1956), it is illustrated graphically in Figure 16. As with Halley's method (5.24), Muller's method is useful for identifying the presence of complex roots.

Polynomials of higher degree are rarely if ever used since, as before, there are two significant problems involved in any such implementation: which root of the polynomial leads to the best approximation and how is the root computed numerically?

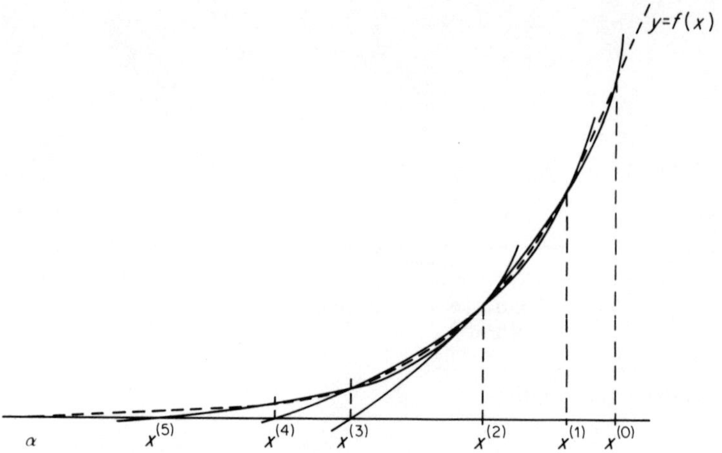

Figure 16 Convergence of Muller's method

(b) Inverse Interpolation

Linear interpolation at $y^{(n)} \equiv f(x^{(n)})$ and $y^{(n-1)} \equiv f(x^{(n-1)})$ leads to

$$P(y) = F^{(n)} + (y - y^{(n)})F_{n,n-1},$$

where $F^{(n)} = F(y^{(n)}) \equiv x^{(n)}$, etc. It follows that

$$x^{(n+1)} \equiv P(0) = x^{(n)} - f^{(n)} \frac{1}{f_{n,n-1}}.$$

Thus there are at least three different formulations leading to the secant method. Quadratic interpolation at $y^{(n)}, y^{(n-1)}$ and $y^{(n-2)}$ leads to an additional term $y^{(n)}y^{(n-1)}F_{n,n-1,n-2}$ from the right-hand side of (6.2). From the definition of divided differences

$$F_{n,n-1,n-2} \equiv \frac{F_{n,n-1} - F_{n-1,n-2}}{y^{(n)} - y^{(n-2)}}$$

$$= \frac{1}{f^{(n)} - f^{(n-2)}} \left\{ \frac{1}{f_{n,n-1}} - \frac{1}{f_{n-1,n-2}} \right\}$$

and so the new formula is

$$x^{(n+1)} = x^{(n)} - \frac{f^{(n)}}{f_{n,n-1}} + \frac{f^{(n)}f^{(n-1)}}{f^{(n)} - f^{(n-2)}} \left\{ \frac{1}{f_{n,n-1}} - \frac{1}{f_{n-1,n-2}} \right\}. \qquad (6.10)$$

Iteration formulae can be generated from higher order polynomials and they have been studied by Traub (1964).

Iteration formulae have also been based on *rational interpolation* rather than polynomial interpolation; that is the solution of $f(x) = 0$ is approximated by a zero of a function of the form

$$R(x) = \frac{a - x}{\sum_{i=1}^{m} b_i x^i}.$$

Such formulae have been studied extensively (Jarratt, 1966 and references therein).

Exercises

6.3. Verify that rational interpolation at $x^{(n)}$, $x^{(n-1)}$ and $x^{(n-2)}$ by a function of the form

$$R(x) = \frac{a - x}{bx + c}$$

leads to the iteration formula

$$x^{(n+1)} = x^{(n)} - f^{(n)} \frac{f^{(n-1)} - f^{(n-2)}}{f_{n,n-2} f^{(n-1)} - f_{n,n-1} f^{(n-2)}}. \tag{6.11}$$

6.4. Verify that *inverse* rational interpolation at $y = f^{(n)}$ and $y = f^{(n-1)}$ by a function of the form

$$R(x) = \frac{1}{a + by}$$

leads to the iteration formula

$$x^{(n+1)} = x^{(n)} - \frac{x^{(n)} f^{(n)}}{f^{(n-1)} + x^{(n)} f_{n,n-1}}.$$

6.3 Order of Convergence

It is possible to derive the order of interpolatory iteration formulae from the expression for the error in interpolation given in (6.3). As with one-point formulae (Section 5.7) it is the inverse formulation that provides the most straightforward construction.

The secant method is derived from linear interpolation at $y = f^{(n)}$ and $y = f^{(n-1)}$; thus it follows from (6.3) that

$$F(y) - P(y) = (y - f^{(n)})(y - f^{(n-1)}) \frac{1}{2} \frac{\mathrm{d}^2 F(\xi)}{\mathrm{d}y^2}.$$

When $y = 0$ and the inverse derivative is replaced by the appropriate form, this becomes

$$\alpha - x^{(n+1)} = \frac{f^{(n)} f^{(n-1)}}{2} \left\{ -\frac{f''(\eta)}{f'(\eta)^3} \right\},$$

where $\eta \equiv F(\xi)$. In addition

$$\lim_{n\to\infty} \frac{f^{(n)}}{(x^{(n)}-\alpha)} \frac{f^{(n-1)}}{(x^{(n-1)}-\alpha)} = f'(\alpha)^2;$$

thus as η tends to α, it follows that

$$\lim_{n\to\infty} \frac{x^{(n+1)}-\alpha}{(x^{(n)}-\alpha)(x^{(n-1)}-\alpha)} = \frac{f''(\alpha)}{2f'(\alpha)}, \qquad f'(\alpha) \neq 0. \tag{6.12}$$

The order of the method (cf. Section 5.5) is the value of p such that

$$\lim_{n\to\infty} \frac{x^{(n+1)}-\alpha}{(x^{(n)}-\alpha)^p} = C\,(\neq 0); \tag{6.13}$$

this is obviously equivalent to

$$\lim_{n\to\infty} \frac{x^{(n)}-\alpha}{(x^{(n-1)}-\alpha)^p} = C. \tag{6.14}$$

It follows by combining (6.13) and (6.14) that another equivalent definition is

$$\lim_{n\to\infty} \frac{x^{(n+1)}-\alpha}{(x^{(n-1)}-\alpha)^{p^2}} = C^{1+p}, \tag{6.15}$$

while from (6.12) and (6.14) it follows that

$$\lim_{n\to\infty} \frac{x^{(n+1)}-\alpha}{(x^{(n-1)}-\alpha)^{1+p}} = \frac{f''(\alpha)}{2f'(\alpha)}\,C. \tag{6.16}$$

Thus by equating the terms in (6.15) and (6.16) it follows that the order is a root of the equation

$$p^2 = 1 + p,$$

that is, $p = \frac{1}{2}(1+\sqrt{5})$; note that the order of an iteration formula is not necessarily an integer. The asymptotic error constant is

$$\left(\frac{f''(\alpha)}{2f'(\alpha)}\right)^{1/p}.$$

Similarly quadratic interpolation at $f^{(n)}, f^{(n-1)}$ and $f^{(n-2)}$ leads to an estimate of the error of the form

$$\alpha - x^{(n+1)} = \frac{f^{(n)}f^{(n-1)}f^{(n-2)}}{6} \frac{\mathrm{d}^3F(\xi)}{\mathrm{d}y^3}$$

for the iteration formula (6.11); by analogy with (6.12) it then follows that

$$\lim_{n\to\infty} \frac{x^{(n+1)}-\alpha}{(x^{(n)}-\alpha)(x^{(n-1)}-\alpha)(x^{(n-2)}-\alpha)} = -\frac{f'(\alpha)^3}{6} \frac{\mathrm{d}^3F(0)}{\mathrm{d}y^3}.$$

As in Section 5.7, iteration formulae derived from direct interpolation can be studied by an analysis of the expression for the interpolation error. Thus it is possible to determine the order of Muller's method, for which the interpolating polynomial is

$$P(x) = f^{(n)} + (x - x^{(n)})f_{n,n-1} + (x - x^{(n)})(x - x^{(n-1)})f_{n,n-1,n-2}.$$

The corresponding interpolation error is

$$f(x) - P(x) \equiv (x - x^{(n)})(x - x^{(n-1)})(x - x^{(n-2)})\,\frac{f'''(\xi)}{3!},$$

where

$$\xi \in \mathrm{Int}\{x, x^{(n)}, x^{(n-1)}, x^{(n-2)}\}.$$

Then

$$P(\alpha) = (\alpha - x^{(n)})(\alpha - x^{(n-1)})(\alpha - x^{(n-2)})\,\frac{f'''(\xi)}{3!},$$

where $\xi \equiv \xi(\alpha)$. The mean value theorem leads to

$$-P(\alpha) = (x^{(n+1)} - \alpha)P'(\eta),$$

where η tends to α as the iteration converges, thus

$$\lim_{n\to\infty} P'(\eta) = \lim_{n\to\infty} f_{n,n-1} = f'(\alpha).$$

Therefore it follows that

$$\lim_{n\to\infty} \frac{x^{(n+1)} - \alpha}{(x^{(n)} - \alpha)(x^{(n-1)} - \alpha)(x^{(n-2)} - \alpha)} = -\frac{1}{6}\,\frac{f'''(\alpha)}{f'(\alpha)},$$

and it follows by analogy with (6.13)–(6.16) that the order of Muller's method is 1.84 (see Exercise 6.5).

Note that direct and inverse interpolation of the same degree lead to iteration formula of the same order – only the asymptotic error constants differ.

Apart from interpolatory iteration formulae, it usually follows that the only way of deriving the order is to expand by Taylor's theorem, each of the individual terms on the right-hand side of the formula

$$x^{(n+1)} = \varphi(x^{(n)}, \ldots, x^{(n-k)}),$$

and then observe which are the dominant terms as the limit is approached.

There is, however, one group of formulae that allow a less tedious analysis, namely formulae that are derived from one-point interpolatory formulae by replacing derivatives by differences. As established in Section 6.1, the simplest such formula defines the secant method, so to avoid undue repetition consider, as an example, the formula (6.5), that is

$$x^{(n+1)} = x^{(n)} - f^{(n)}\left\{\frac{1}{f_{n,n-1}} + \frac{1}{f_{n,n-2}} - \frac{1}{f_{n-1,n-2}}\right\}. \tag{6.17}$$

This formula is derived via the inverse formulation; thus the differences correspond to an approximation of $\mathrm{d}F(y^{(n)})/\mathrm{d}y$.

If the inverse function is interpolated at $y^{(n)}$, $y^{(n-1)}$ and $y^{(n-2)}$ by a polynomial $Q(y)$, it follows from (6.2) that

$$\frac{\mathrm{d}Q(y^{(n)})}{\mathrm{d}y} = F_{n,n-1} + (y^{(n)} - y^{(n-1)})F_{n,n-1,n-2}$$

$$\equiv \frac{1}{f_{n,n-1}} + \frac{1}{f_{n,n-2}} - \frac{1}{f_{n-1,n-2}}$$

(see Exercise 6.9). If $P(y)$ is a linear polynomial that interpolates $F(y)$ and $\mathrm{d}F(y)/\mathrm{d}y$ at $y = y^{(n)}$, it follows from the inverse formulation of Newton's method that the iteration (6.17) can be written as

$$x^{(n+1)} = P(0) + f^{(n)}\left\{\frac{\mathrm{d}F(y^{(n)})}{\mathrm{d}y} - \frac{\mathrm{d}Q(y^{(n)})}{\mathrm{d}y}\right\}$$

It follows by differentiating (6.3) that

$$\frac{\mathrm{d}F(y^{(n)})}{\mathrm{d}y} - \frac{\mathrm{d}Q(y^{(n)})}{\mathrm{d}y} = (f^{(n)} - f^{(n-1)})(f^{(n)} - f^{(n-2)})\frac{1}{6}\frac{\mathrm{d}^3F(\xi)}{\mathrm{d}y^3}$$

(see Exercise 6.7). It is shown in Section 5.7 that

$$F(0) - P(0) = \frac{{f^{(n)}}^2}{2}\frac{\mathrm{d}^2F(\eta)}{\mathrm{d}y^2},$$

thus

$$\alpha - x^{(n+1)} = \frac{{f^{(n)}}^2}{2}\frac{\mathrm{d}^2F(\eta)}{\mathrm{d}y^2} - f^{(n)}(f^{(n)} - f^{(n-1)})(f^{(n)} - f^{(n-2)})\frac{1}{6}\frac{\mathrm{d}^3F(\xi)}{\mathrm{d}y^3}.$$

It can be shown (Traub, 1964) that, as the iteration converges, the second term dominates the right-hand side, which is then asymptotically equal to

$$(x^{(n)} - \alpha)(x^{(n-1)} - \alpha)(x^{(n-2)} - \alpha)\frac{1}{6}\frac{\mathrm{d}^3F(0)}{\mathrm{d}y^3};$$

thus the convergence is similar to that for the iteration formulae derived from quadratic interpolation – for example, Muller's method.

Exercises

6.5. Verify that the order p of the iteration formula (6.11) satisfies

$$p^3 = 1 + p + p^2,$$

i.e. $p \approx 1.84$. Determine the form of the asymptotic error constant.

6.6. Prove that inverse interpolation at m points ($m \geqslant 2$) leads to an iteration formula of order p, where

$$p^m = \sum_{i=0}^{m-1} p^i.$$

6.7. Prove that if $Q(y)$ is a polynomial of degree at most k that interpolates $F(y)$ at $y^{(n)}, \ldots, y^{(n-k)}$, and if $F(y)$ is sufficiently differentiable, then there exists a point ξ_n such that

$$\frac{\mathrm{d}F(y^{(n)})}{\mathrm{d}y} - \frac{\mathrm{d}Q(y^{(n)})}{\mathrm{d}y} = \prod_{l=1}^{k} (y^{(n)} - y^{(n-l)}) \frac{\mathrm{d}^{k+1}F(\xi_n)}{\mathrm{d}y^{k+1}}.$$

6.8. Verify that the formulae (6.4) and (6.5) have the same order.

6.9. Prove that

$$(y^{(n)} - y^{(n-1)})F_{n,n-1,n-2} = F_{n,n-2} - F_{n-1,n-2} = \frac{1}{f_{n,n-2}} - \frac{1}{f_{n-1,n-2}}.$$

6.10. Assume that $f'(x)$ and $f''(x)$ are positive in some interval I that contains a root α. Use the mean value theorem to prove that for any $x^{(0)} > x^{(1)} > \alpha$ such that $x^{(0)}, x^{(1)} \in I$, the sequence of secant iterates is monotonically decreasing and it converges to α.

6.4 Bracketing Methods

Cautionary example The Newton method and the secant method are both applied to the equation

$$\cos\left(\frac{x^2 + 5}{x^4 + 1}\right) = 0;$$

it is known that this equation has a single root in the interval [0, 1]. The results of the iterations, using the ends of the interval as the initial approximations, are given in Table 7.

What would happen if when Newton's method was started with $x^{(0)} = 0.0$?

Second cautionary example Newton's method is applied to the equation

$$e^x - 1.0 - \cos(\pi x) = 0;$$

it is known that this equation has a single positive root which lies in the interval [0, 1]. The iteration is started at each end of the interval. In addition it is started

Table 7 Divergence of secant and Newton iterations

secant	n	Newton
0.0	0	1.0
1.0	1	2.40
0.2227	2	61.5
0.4122	3	4.4 E 8
5.37	4	
−1.79	5	
−23.2	6	
61.7	7	
−4.9 E 7	8	

with the initial approximations $-\epsilon$ and $1-\epsilon$, where ϵ is the smallest value such that, using floating point arithmetic, $1+\epsilon > 1$; it follows that the number is computer dependent. A C.T.L. modular one computer was used to generate the results given in Table 8, and for such a machine $\epsilon = 1.86\ \text{E} - 9$.

What would happen if the calculation were repeated with $x^{(0)} = 1.0$, using *exact arithmetic*?

One class of methods, designed to avoid the computational pitfalls illustrated in the above examples, is the group of so-called *bracketing methods.* Generally, a bracketing method is an iterative procedure that requires an interval $[a, b]$ such

Table 8 Instability of Newton's Method

n	x			
0	0.0	1.0	$-\epsilon$	$1.0-\epsilon$
1	0.99999999	−1.86 E −9	1.00000001	4.66 E −9
2	0.00000002	1.00000001	−0.00000009	0.99999995
3	0.99999982	−0.00000009	1.0000009	0.00000014
4	0.00000048	1.0000009	−0.00000249	0.99999852
5	0.99999482	−0.00000249	1.0000270	0.00000388
6	0.00001363	1.0000270		0.99995785
7	0.99985188	−0.0000711	as for	0.00011088
8	0.00038950	1.0007740		0.99879631
9	0.99578491	−0.0063877	$x^{(0)} = 1.0$	0.00315550
10	0.01095264	3.2763161		0.96688652
11	0.89416622	2.1914516		0.07907769
12	0.20539062	1.5272919		0.55731518
13	0.38853989	−0.8581568		0.36605818
14	0.35876533	−0.5073535		0.35827160
15	0.35823240	−0.6549953		0.35823221
16	0.35823221	−0.6605951		
17		−0.6606242		
18		−0.6606242		

that $a \leqslant \alpha \leqslant b$ at the start of each step. At the end of the step, the procedure provides a smaller sub-interval $[a^+, b^-]$ such that $a \leqslant a^+ \leqslant \alpha \leqslant b^- \leqslant b$ and $b - a > b^- - a^+$. A condition invariably imposed on the interval is that $f(a)f(b) < 0$; thus if $f(x)$ is a continuous function the presence of at least one root in the interval is guaranteed. During the course of the iteration, at least one of the sequences of end-points of the intervals converges to a root, in most bracketing iterations both end-points converge to the same root and hence the length of the intervals tends to zero. An example of bracketing method in which this occurs is the method of *bisection*, and one method in which it frequently does not is *regula falsi.*

Bisection

The algorithm is a procedure for successively halving the size of the interval, ensuring that there is still a sign change between the end-points.

Algorithm If $f(a)f(b) \geqslant 0$ the method cannot proceed; otherwise at each step:

$$\text{if } f\left(\frac{a+b}{2}\right)f(a)\begin{cases} >0 & \text{replace } a \text{ by } \dfrac{a+b}{2} \\[2ex] <0 & \text{replace } b \text{ by } \dfrac{a+b}{2} \\[2ex] =0 & \dfrac{a+b}{2} \text{ is the required solution, stop.} \end{cases}$$

This method is always slow to converge, but convergence is guaranteed.

Since the number of steps necessary to obtain an interval of a particular size depends solely on the value of $(b - a)$ and not on the form of the function $f(x)$, it is possible to specify in advance the number of iterations necessary. After k iterations, the root has been restricted to a sub-interval of length $(b - a)2^{-k}$.

Regula falsi (The method of false position)

The iteration is similar to the secant method, but now the two points used to define the linear interpolation must bracket the solution.

Algorithm If $f(x^{(0)})f(y^{(0)}) \geqslant 0$ the method cannot proceed, otherwise for $n = 0, 1, \ldots, f^{(n)} = f(x^{(n)}), g^{(n)} = f(y^{(n)})$

$$x^{(n+1)} = x^{(n)} - \frac{x^{(n)} - y^{(n)}}{f^{(n)} - g^{(n)}} f^{(n)}, \tag{6.17}$$

$$\text{if } f(x^{(n+1)})f(x^{(n)})\begin{cases} <0 & y^{(n+1)} = x^{(n)} \\ >0 & y^{(n+1)} = y^{(n)} \\ =0 & x^{(n+1)} \text{ is the required solution.} \end{cases}$$

The terminating condition for this iteration should depend on $|x^{(n)} - x^{(n+1)}|$, $|x^{(n)} - y^{(n)}|$, $|f^{(n)}|$ or n, and an example of a suitable condition is given in the Pegasus algorithm.

The convergence of the regula falsi method is illustrated in Figure 17; this show the major drawback of the method for functions that are concave or convex in the neighbourhood of the solution. The convergence is slow and from one side only; thus the length of the interval does not tend to zero as the root is approached. It can be seen from (6.17) and (6.9) that when $y^{(n)}$ is constant and does not converge to the solution α, it follows that

$$\lim_{n\to\infty} \frac{x^{(n+1)} - \alpha}{x^{(n)} - \alpha} = (y - \alpha)\frac{1}{2}\frac{f''(\alpha)}{f'(\alpha)},$$

where $y^{(n)} = y$, for sufficiently large n. Hence the convergence is ultimately linear.

Numerical example (due to Dowell and Jarratt (1971)) The regula falsi method was used to compute a root of

$$x^2 - (1 - x)^5 = 0$$

in the interval $[0, 1]$. The results of the iteration are given in Table 9. Note that, if it is assumed that $\alpha = 0.345\,954\,82$, then

$$\frac{x^{(n+1)} - \alpha}{x^{(n)} - \alpha} \simeq 0.444, \qquad n = 3, 4, \ldots.$$

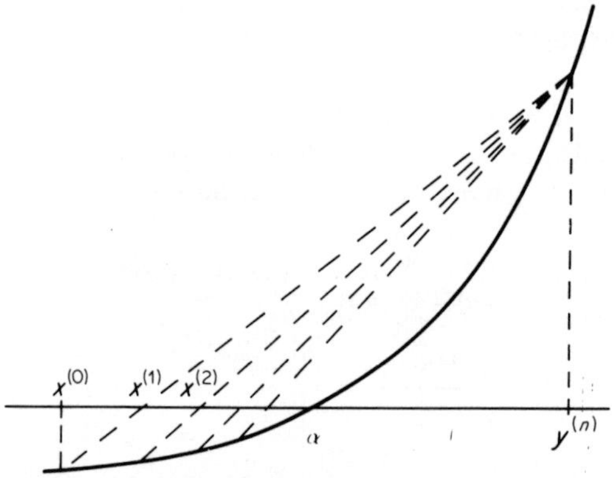

Figure 17 Convergence of the regula falsi method

Table 9 Regula falsi iteration

$10^8(x-\alpha)$	x	y
	1.0000000	0.0
	0.50000000	0.0
	0.41025641	0.0
	0.37398949	0.0
	0.35832109	0.0
547679	0.35143161	0.0
242939	0.34838421	0.0
107833	0.34703315	0.0
47877	0.34643359	0.0
21260	0.34616742	0.0
19441	0.34604923	0.0
4192	0.34599674	0.0
1862	0.34597344	0.0
826	0.34596308	0.0
367	0.34595849	0.0
163	0.34595645	0.0
72	0.34595554	0.0
32	0.34595514	0.0
14	0.34595496	0.0
6	0.34595488	0.0
2	0.34595484	0.0
1	0.34595483	0.0
	0.34595482	0.0
	0.34595482	

Pegasus

Several modifications of the basic regula falsi method have been suggested, one of the most straightforward being the Pegasus method (Dowell and Jarratt, 1972); a similar and marginally less efficient method, the Illinois method (Dowell and Jarratt, 1971), will not be discussed.

The strategy behind the Pegasus method is to modify the regula falsi step (6.17) if convergence appears to be linear; that is if $f^{(n+1)}f^{(n)} > 0$ and hence $y^{(n+1)} = y^{(n)}$. The numerical value of $g^{(n+1)}$ is then taken to be

$$g^{(n+1)} = g^{(n)} \frac{f^{(n)}}{f^{(n)} + f^{(n+1)}}\,.$$

With this modification convergence is invariably much faster and from both sides, i.e. $\lim_{n\to\infty} |x_n - y_n| = 0$. But examples can be constructed where this does not hold (see Exercise 6.13). The algorithm for this method is presented in the form of an

ALGOL 68 procedure:

```
'PROC' PEGSOLVE = ('REF''REAL' XN,YN,'REAL' DEL1,DEL2,'INT' ITMAX,
                  'REF''BOOL' BRACKET, 'PROC'('REAL')'REAL' F)'VOID'
'BEGIN' 'REAL' GN:=F(YN), FN:=F(XN), XNPLUS1, FNPLUS1;
'IF' BRACKET:=(FN*GN < Ø)
'THEN'
    'TO' ITMAX 'WHILE' XNPLUS1:=XN-FN*(XN-YN)/(FN-GN);
                       FNPLUS1:=F(XNPLUS1);
                       'ABS'(YN-XN)>DEL1 'AND' 'ABS'FNPLUS1 >DEL2
    'DO'
        'IF' FN*FNPLUS1>Ø 'THEN' GN*:=FN/(FN+FNPLUS1)
                          'ELSE' YN:=XN; GN:=FN 'FI';
        XN:=XNPLUS1; FN:=FNPLUS1;
        PRINT ((XN,YN,FN,GN,NEWLINE))
    'OD'
'ELSE'
    PRINT (("THERE IS NO SOLUTION BETWEEN ",X," AND ",XS,NEWLINE))
'FI'
'END';
```

Numerical example The Pegasus method was used to find a root of

$$x^2 - (1 - x)^5 = 0$$

and the results are given in Table 10.

Table 10 The Pegasus method

n	x	y
0	1.00000000	0.00000000
1	0.50000000	0.00000000
2	0.39475713	0.00000000
3	0.35184685	0.00000000
4	0.34585928	0.35184685
5	0.34595544	0.34585928
6	0.34595482	0.34595544
7	0.34595482	0.34595544

Other bracketing methods have been put forward at various times (for example Bus and Dekker, 1975 and Cox, 1970) these invariably involve some combination of bisection and local interpolation – some authors favour polynomial interpolation, while others prefer to use rational interpolation. The strategy in such *poly-algorithms* is to perform interpolation with the sequence $\{x^{(n)}\}$ and to accept the value of $x^{(n+1)}$ this interpolation provides only if it lies in an interval defined by $x^{(n)}$ and $y^{(n)}$ – otherwise a bisection step is used to define $x^{(n+1)}$.

In 1824, G.P. Dandelin (see Ostrowski, 1966) put forward a method which is a hybrid of the Newton and secant methods and which, for convex or concave functions, provides a method of bracketing the solution. It is similar in spirit to the Fourier method given in Section 5.10, but in this method the Newton iterates $\{x^{(n)}\}$ are used to derive a second sequence $\{y^{(n)}\}$ such that

$$y^{(n+1)} = y^{(n)} - \frac{y^{(n)} - x^{(n)}}{g^{(n)} - f^{(n)}} g^{(n)}, \qquad n = 1, 2, \ldots, \tag{6.18}$$

where $g^{(n)}$ and $f^{(n)}$ are as in (6.17). When the function $f(x)$ is convex or concave, $x^{(n)}$ and $y^{(n)}$ bracket the solution.

Exercises

6.11. Assume that $f'(x)$ and $f''(x)$ are positive in some interval I which contains a root α. Use the mean value theorem to prove that if $x^{(0)} > \alpha > y^{(0)}$ and $x^{(0)}$, $y^{(0)} \in I$, then $\{y^{(n)}\}$ defined by (6.18) is a monotonically increasing sequence that converges to α.

6.12. Subject to the same conditions on $f(x)$ as above and with $\{y^{(n)}\}$ defined by (6.18), where $\{x^{(n)}\}$ is a sequence of secant iterates, use the mean value theorem to prove that if $x^{(0)} > x^{(1)} > \alpha > y^{(0)}$ and $x^{(0)}, y^{(0)} \in I$, then $\{y^{(n)}\}$ converges monotonically to α.

6.13. Use the Pegasus method to find a root of $f(x) = 0$ in the interval $[-1, 1]$ when

$$f(x) \equiv \begin{cases} 2x & x < 0 \\ x^2 & x \geqslant 0. \end{cases}$$

Note that the Illinois method performs significantly better on this example.

7

Systems of Equations

In Section 5.7 the Newton method for systems of equations is introduced; frequently however it is not feasible to recompute the components of the Jacobian matrix and solve

$$J(\mathbf{x}^{(n)})\delta^{(n)} = -\mathbf{f}(\mathbf{x}^{(n)})$$

at each iteration. A few suggestions for reducing the amount of computation are also given, but they invariably replace one drawback with another such as that $J(\mathbf{x})$ should be diagonally dominant, the convergence is slow or a good approximation to $J(\alpha)$ is required.

7.1 The Secant Method

An alternative approach is to replace the derivatives in the Jacobian by differences. Probably the most straightforward replacement is

$$\frac{\partial f_i}{\partial x_j} \approx \frac{f_i(\mathbf{x}^{(n)} + \delta_j \mathbf{e}_j) - f_i(\mathbf{x}^{(n)})}{\delta_j}, \qquad i, j = 1, \ldots, N,$$

where $\mathbf{e}_j$ are the unit vectors. The most obvious drawbacks with such an approach are that it still requires $N^2 + N$ function evaluations at each step and a particular choice of the constants δ_j; some work has been done on this latter problem, for example by Broyden (1965) and by Curtis and Reid (1972). An alternative approach is to develop an N-dimensional analogue of the *secant method*, based on linear interpolation of each component of $\mathbf{f}(\mathbf{x})$.

Consider, as an example, the solution of the system $f(x, y) = 0, g(x, y) = 0$; the iteration can be based on bi-variate linear interpolation at $\mathbf{x} = \mathbf{x}^{(n)}, \mathbf{x}^{(n-1)}$ and $\mathbf{x}^{(n-2)}$. The polynomial that interpolates $f(x, y)$ can be written as

$$P(x, y) \equiv f^{(n)} + \frac{x - x^{(n)}}{d} \det \begin{bmatrix} f^{(n-1)} - f^{(n)} & f^{(n-2)} - f^{(n)} \\ y^{(n-1)} - y^{(n)} & y^{(n-2)} - y^{(n)} \end{bmatrix}$$

$$+ \frac{y - y^{(n)}}{d} \det \begin{bmatrix} x^{(n-1)} - x^{(n)} & x^{(n-2)} - x^{(n)} \\ f^{(n-1)} - f^{(n)} & f^{(n-2)} - f^{(n)} \end{bmatrix}, \qquad (7.1$$

where

$$d \equiv \det \begin{bmatrix} x^{(n-1)} - x^{(n)} & x^{(n-2)} - x^{(n)} \\ y^{(n-1)} - y^{(n)} & y^{(n-2)} - y^{(n)} \end{bmatrix}$$

and $f^{(n)} \equiv f(x^{(n)}, y^{(n)})$, etc. If the equivalent approximation of $g(x, y)$ is $Q(x, y)$, the next iterate is the solution of the linear system

$$P(x^{(n+1)}, y^{(n+1)}) = 0, \qquad Q(x^{(n+1)}, y^{(n+1)}) = 0.$$

It follows from (7.1) and the corresponding expression for $Q(x, y)$ that

$$\boldsymbol{\delta}^{(n)} \equiv \mathbf{x}^{(n+1)} - \mathbf{x}^{(n)} \tag{7.2}$$

is given by

$$J^{(n)} \boldsymbol{\delta}^{(n)} = -\mathbf{f}(\mathbf{x}^{(n)}), \tag{7.3}$$

where the components of $J^{(n)}$ – an approximation to $J(\mathbf{x}^{(n)})$ – are ratios of determinants.

Two significant problems with such a formulation of the secant method in N dimensions are the need for $N + 1$ *starting approximations* and a matrix $J^{(n)}$ that has to be evaluated at each iteration. Both these difficulties are overcome in the *Barnes secant method* (Barnes, 1965). It follows from (7.1) etc. that for a general system of equations $\mathbf{f}(\mathbf{x}) = 0$,

$$\mathbf{f}^{(n+1)} = \mathbf{f}^{(n)} + J^{(n+1)} \boldsymbol{\delta}^{(n)}, \tag{7.4}$$

where $\mathbf{f}^{(n)} \equiv \mathbf{f}(\mathbf{x}^{(n)})$, etc. It is possible to use this equation, together with (7.2) and (7.3), to determine a matrix $J^{(n+1)}$ in the form

$$J^{(n+1)} \equiv J^{(n)} + D^{(n)}. \tag{7.5}$$

Combining (7.3), (7.4) and (7.5) leads to

$$\mathbf{f}^{(n+1)} = D^{(n)} \boldsymbol{\delta}^{(n)},$$

which for given $\mathbf{x}^{(n+1)}$ and $\mathbf{x}^{(n)}$ is satisfied by any matrix of the form

$$D^{(n)} = \frac{\mathbf{f}^{(n+1)} \mathbf{z}^{(n)\mathrm{T}}}{\mathbf{z}^{(n)\mathrm{T}} \boldsymbol{\delta}^{(n)}} \tag{7.6}$$

for arbitrary $\mathbf{z}^{(n)}$. The Barnes algorithm – given $\mathbf{x}^{(0)}$ and $J^{(0)}$ – consists therefore of (7.3) and (7.6), together with a strategy for selecting the vectors $\mathbf{z}^{(n)}$ $(n = 0, 1, \ldots)$ that ensures the method is equivalent to the secant method formulated in terms of interpolating polynomials.

The strategy suggested by Barnes is that

$$\mathbf{z}^{(n)\mathrm{T}} \boldsymbol{\delta}^{(k)} = 0, \quad k = \begin{cases} 0, \ldots, n-1, & n < N \\ n - N + 1, \ldots, n-1 & n \geqslant N. \end{cases}$$

In any implementation of the method this orthogonalization would be accomplish using the *modified Gram–Schmidt* process (Rice, 1966). It follows from (7.6) that for this choice of the sequence $\mathbf{z}^{(n)}$

$$D^{(j)}\delta^{(k)} = 0, \qquad 0 < j - k < N,$$

and so from (7.5)

$$\begin{aligned} J^{(n)}\delta^{(n-i)} &= \{J^{(n-i+1)} + D^{(n-i+1)} + \cdots + D^{(n-1)}\}\delta^{(n-i)} \\ &= J^{(n-i+1)}\delta^{(n-i)}, \qquad i = 1, \ldots, N, \end{aligned} \tag{7.7}$$

and then from (7.4),

$$J^{(n)}\delta^{(n-i)} = \mathbf{f}^{(n-i+1)} - \mathbf{f}^{(n-i)}.$$

Hence

$$\begin{aligned} J^{(n)}(\mathbf{x}^{(n)} - \mathbf{x}^{(n-i)}) &= J^{(n)}(\delta^{(n-1)} + \cdots + \delta^{(n-i-1)}) \\ &= \mathbf{f}^{(n)} - \mathbf{f}^{(n-i)}, \qquad i = 1, \ldots, N; \end{aligned}$$

thus

$$\mathbf{P}(\mathbf{x}) \equiv \mathbf{f}(\mathbf{x}^{(n)}) + J^{(n)}(\mathbf{x} - \mathbf{x}^{(n)})$$

is a linear system that interpolates $\mathbf{f}(\mathbf{x})$ at the points $\mathbf{x} = \mathbf{x}^{(n-i)}$ $(i = 0, \ldots, N)$, an the equivalence of the two formulations of the secant method is verified, provided the directions $\delta^{(n-i)}$ are linearly independent.

Householder's formula (alternatively known as the Sherman–Morrison–Woodb formula) states that if

$$B = A + \mathbf{u}\mathbf{w}^{\mathrm{T}},$$

then

$$B^{-1} = A^{-1} - \frac{A^{-1}\mathbf{u}\mathbf{w}^{\mathrm{T}}A^{-1}}{1 + \mathbf{w}^{\mathrm{T}}A^{-1}\mathbf{u}}.$$

It follows from Householder's formula that it is possible, starting with an initial approximation $H^{(0)} \equiv [J^{(0)}]^{-1}$, to update the approximate inverse Jacobian $H^{(n)} \equiv [J^{(n)}]^{-1}$ such that

$$H^{(n+1)} = H^{(n)} + C^{(n)}, \qquad n = 0, 1, \ldots.$$

Then it is no longer necessary to solve the linear system (7.3), as it has been replaced by the matrix multiplication

$$\delta^{(n)} = -H^{(n)}\mathbf{f}^{(n)}. \tag{7.8}$$

In the secant method it follows that as $D^{(n)} = \mathbf{u}\mathbf{w}^{\mathrm{T}}$, where

$$\mathbf{u} = \mathbf{f}^{(n+1)} \qquad \text{and} \qquad \mathbf{w} = \frac{\mathbf{z}^{(n)}}{\mathbf{z}^{(n)\mathrm{T}}\,\delta^{(n)}},$$

ouseholder's formula leads to the inverse modification

$$C^{(n)} = -\frac{H^{(n)}\mathbf{f}^{(n+1)}\mathbf{z}^{(n)\mathrm{T}}H^{(n)}}{\mathbf{z}^{(n)\mathrm{T}}\boldsymbol{\delta}^{(n)} + \mathbf{z}^{(n)\mathrm{T}}H^{(n)}\mathbf{f}^{(n+1)}}. \tag{7.9}$$

more efficient method of implementing these modification methods is to store ıe *matrix factors of* $J^{(n)}$ rather than its inverse, and then to update the factors. his form of implementation is considered in Section 7.3(a).

Exercises

.1. Verify that (7.4) is valid for the secant method defined in two dimensions by 7.1), etc.

.2. Assume that for some $m \geqslant 1$, there exist constants a_j $(j = 1, \ldots, m)$ such that

$$\boldsymbol{\delta}^{(m+1)} = a_1\boldsymbol{\delta}^{(1)} + \cdots + a_m\boldsymbol{\delta}^{(m)}; \tag{7.10}$$

rove that (7.3) and (7.4) lead to

$$-\mathbf{f}^{(m+1)} = \sum_{j=1}^{m} a_j\,(\mathbf{f}^{(j+1)} - \mathbf{f}^{(j)}). \tag{7.11}$$

ence prove that if $\mathbf{f}(\mathbf{x}) = \mathbf{b} + J\mathbf{x}$, then it follows from (7.4) and (7.11) that $^{m+2)} = 0$. (Thus as it is clear from (7.10) that $m \leqslant N$, convergence to the root f a linear system requires no more than $N + 2$ iterations – ignoring round-off rrors.)

.3. Verify the Householder modification formula.

.4. Verify that alternative equivalent forms of (7.9) are

$$C^{(n)} = -\frac{H^{(n)}\mathbf{f}^{(n+1)}\mathbf{z}^{(n)\mathrm{T}}H^{(n)}}{\mathbf{z}^{(n)\mathrm{T}}H^{(n)}\boldsymbol{\Delta}^{(n)}} = \frac{(\boldsymbol{\delta}^{(n)} - H^{(n)}\boldsymbol{\Delta}^{(n)})\mathbf{z}^{(n)\mathrm{T}}H^{(n)}}{\mathbf{z}^{(n)\mathrm{T}}H^{(n)}\boldsymbol{\Delta}^{(n)}} \tag{7.12}$$

here $\boldsymbol{\Delta}^{(n)} \equiv \mathbf{f}^{(n+1)} - \mathbf{f}^{(n)}$.

7.2 Quasi-Newton Methods

In general, methods that approximate the Jacobian in some way and then use ıat approximation in a formula such as (7.3) or (7.8) are called *modified Newton ıethods*; methods in which updating matrices $C^{(n)}$ and $D^{(n)}$ are matrices of rank ne are called *rank-one modification* methods. The particular class of methods that ıtisfy a set of additional conditions – in the Barnes secant method this is the ystem (7.4) – are called *quasi-Newton methods.*

The secant method determines the correction to the approximate solution from 7.3); in a general quasi-Newton method it follows that

$$\boldsymbol{\delta}^{(n)} = \sigma_n\mathbf{p}^{(n)},$$

here

$$\mathbf{p}^{(n)} = -H^{(n)}\mathbf{f}(\mathbf{x}^{(n)});$$

the sequence of scalars $\{\sigma_n\}$ varies from method to method. For a large class of methods the additional condition (7.4) is used, but now written as

$$H^{(n+1)}\Delta^{(n)} = \sigma_n \mathbf{p}^{(n)}. \tag{7.13}$$

A more general form has been discussed (Broyden, 1965) but it is rarely implemented in practice. It follows from (7.13) that the modification $C^{(n)} \equiv H^{(n+1)} - H^{(n)}$ satisfies

$$C^{(n)}\Delta^{(n)} = \sigma_n \mathbf{p}^{(n)} - H^{(n)}\Delta^{(n)}. \tag{7.14}$$

This condition is satisfied by any matrix of the form

$$C^{(n)} = \frac{(\sigma_n \mathbf{p}^{(n)} - H^{(n)}\Delta^{(n)})\mathbf{v}^{(n)\mathrm{T}}}{\mathbf{v}^{(n)\mathrm{T}}\Delta^{(n)}}, \tag{7.15}$$

which is similar to the correction matrix (7.12) for the secant method. Broyden suggested that a suitable condition for specifying the sequence $\{\mathbf{v}^{(n)}\}$ is that

$$(J^{(n+1)} - J^{(n)})\mathbf{q} = 0 \tag{7.16}$$

for all $\mathbf{q}$ such that

$$\mathbf{p}^{(n)\mathrm{T}}\mathbf{q} = 0. \tag{7.17}$$

This leads to

$$\mathbf{v}^{(n)} = H^{(n)\mathrm{T}}\mathbf{p}^{(n)}, \qquad n = 0, 1, \ldots \tag{7.18}$$

(see Exercise 7.5). With this choice of $\{\mathbf{v}^{(n)}\}$, it follows that the convergence proof for linear problems given for the Barnes method remains valid only if

$$\delta^{(n)\mathrm{T}}\delta^{(k)} = 0, \qquad 0 < n - k < N,$$

which will not be true in general. This apparent defect does not however prevent the method from performing very effectively on non-linear problems (Broyden, 1965, 1968). The convergence has been studied by other methods (Dennis, 1971).

The parameters σ_n ($n = 0, 1, \ldots$) are used to ensure that the method is *norm reducing*. It is important to note that although this strategy prevents divergence, it *does not ensure convergence.* It is found that it is not necessary to minimize the norm; that is, select any σ_n such that

$$\| \mathbf{f}(\mathbf{x}^{(n)} + \sigma_n \mathbf{p}^{(n)}) \| < \| \mathbf{f}(\mathbf{x}^{(n)}) \|$$

and not spend time searching for

$$\min_{\sigma} \| \mathbf{f}(\mathbf{x}^{(n)} + \sigma \mathbf{p}^{(n)}) \|.$$

Although the minimizing strategy provides the greatest local improvement, it is often ultimately slower and invariably more expensive in computer time. A suitable strategy – see for example Broyden (1965) – is first to try $\sigma_n = 1$ and to try other alues only when this does not decrease the norm. It is usual to use the l_2-norm

$$\| \mathbf{f} \|_2 \equiv \{\mathbf{f}^{\mathrm{T}}\mathbf{f}\}^{1/2}.$$

Other quasi-Newton methods have been proposed, mostly for the problems in hich the Jacobian is symmetric, as such methods can then be efficiently applied o optimization problems – that is finding maxima or minima. The form (7.16) is ot the most general and it is easy enough to verify that it is possible to use

$$C^{(n)} = \frac{\sigma_n \mathbf{p}^{(n)}\mathbf{u}^{(n)\mathrm{T}}M}{\mathbf{u}^{(n)\mathrm{T}}M\Delta^{(n)}} - \frac{H^{(n)}\Delta^{(n)}\mathbf{v}^{(n)\mathrm{T}}M}{\mathbf{v}^{(n)\mathrm{T}}M\Delta^{(n)}} \qquad \text{(a rank two update)}$$

or any matrix M and two arbitrary sequences $\{\mathbf{u}^{(n)}\}$ and $\{\mathbf{v}^{(n)}\}$ and still satisfy 7.15). In order to maintain the symmetry of the approximate Jacobian, take

$$\mathbf{u}^{(n)\mathrm{T}}M = \mathbf{p}^{(n)\mathrm{T}}$$

nd

$$\mathbf{v}^{(n)\mathrm{T}}M = \Delta^{(n)\mathrm{T}}H^{(n)};$$

hen

$$C^{(n)} = \frac{\sigma_n}{\mathbf{p}^{(n)\mathrm{T}}\Delta^{(n)}}\,\mathbf{p}^{(n)}\mathbf{p}^{(n)\mathrm{T}} - \frac{H^{(n)}\Delta^{(n)}\Delta^{(n)\mathrm{T}}H^{(n)}}{\Delta^{(n)\mathrm{T}}H^{(n)}\Delta^{(n)}}.$$

This leads to the well-known *Davidon–Fletcher–Powell* (D–F–P) method (Fletcher nd Powell, 1963). Quasi-Newton methods used in optimization are frequently enamed *variable metric methods.* A considerable literature exists comparing the erformance of extant methods and suggesting improvements – for example Fletcher (1970) or more recently Rheinboldt (1974).

Exercises

7.5.(a) Verify using Householder's formula that if

$$H^{(n+1)} \equiv H^{(n)} + C^{(n)}$$

and $C^{(n)}$ is given by (7.15), it follows that

$$J^{(n+1)} = J^{(n)} - \frac{(\Delta^{(n)} + \sigma_n \mathbf{f}^{(n)})\mathbf{v}^{(n)\mathrm{T}}J^{(n)}}{\sigma_n \mathbf{v}^{(n)\mathrm{T}}\mathbf{f}^{(n)}}, \tag{7.19}$$

and hence that if $\mathbf{v}^{(n)\mathrm{T}} = \mathbf{p}^{(n)\mathrm{T}}H^{(n)}$, then

$$J^{(n+1)} = J^{(n)} + \frac{(\Delta^{(n)} + \sigma_n \mathbf{f}^{(n)})\mathbf{p}^{(n)\mathrm{T}}}{\sigma_n \mathbf{p}^{(n)\mathrm{T}}\mathbf{p}^{(n)}}. \tag{7.20}$$

(Note that this form of update contains vector–vector multiplications only, no matrix–vector multiplications.)

(b) Verify, using Householder's formula, that (7.15), (7.16) and (7.18) lead to (7.17).

7.6. What is the update matrix $C^{(n)}$ that corresponds to

$$D^{(n)} = \frac{(\Delta^{(n)} + \sigma_n \mathbf{f}^{(n)})(\Delta^{(n)} + \sigma_n \mathbf{f}^{(n)})^{\mathrm{T}}}{\sigma_n(\Delta^{(n)} + \sigma_n \mathbf{f}^{(n)\mathrm{T}} \mathbf{p}^{(n)})}?$$

7.3 Efficient Implementation of quasi-Newton Methods

The updating of the sequence $\{H^{(n)}\}$ can lead to problems with the build up of round-off errors in the calculation. As a result of the build up, it often follows that:

(1) the computed matrix can have components that are significantly different from the corresponding theoretical values;
(2) the calculation may cause overflow on the machine, even though the components of $H^{(n+1)}$ are not very large;
(3) if $H^{(n)}$ is *nearly* singular all subsequent $H^{(n+k)}$ $(k = 1, 2, \ldots)$ tend to remain nearly singular.

Thus there are sound computational reasons why, on the grounds of accuracy, any implementation of a quasi-Newton method should update $J^{(n)}$ rather than $H^{(n)}$.

Since there are $\frac{2}{3}N^3 + O(N^2)$ arithmetic operations involved in computing the solution of the system of N linear equations defined by

$$J^{(n)}\mathbf{p}^{(n)} = -\mathbf{f}^{(n)},$$

whereas there are $2N^2 + O(N)$ arithmetic operations involved in evaluating the matrix–vector inner product

$$\mathbf{p}^{(n)} = -H^{(n)}\mathbf{f}^{(n)}.$$

It is therefore vitally important to use a numerical method that reduces the arithmetic if updating the sequence $\{J^{(n)}\}$ is to be as rapid as updating the sequence $\{H^{(n)}\}$, and two such methods are described below: (a) updating the orthogonal factors of $J^{(n)}$, and (b) modifying the correction $D^{(n)}$ to use the known structure of the matrix $J^{(n)}$.

(a) Updating the orthogonal factors

This method, due to Gill and Murray (1972), assumes that $J^{(n)}$ is not stored explicitly but is stored in factored form as $L^{(n)}Q^{(n)}$. That is, the lower triangular matrix $L^{(n)}$ and the orthogonal matrix $Q^{(n)}$ are stored separately. The object of the method is therefore as follows: given $L^{(n)}$, $Q^{(n)}$ and $D^{(n)}$, to determine as efficiently as possible the matrices $L^{(n+1)}$ and $Q^{(n+1)}$ such that

$$J^{(n+1)} = J^{(n)} + D^{(n)} = L^{(n+1)}Q^{(n+1)}.$$

A discussion of the existence and uniqueness of the orthogonal factors of a matrix is found in Section 3.1. The use of the orthogonal factors is preferred to that of the triangular factors in this situation, as the latter can lead to numerical instabilities.

It is assumed that vectors $\mathbf{u}^{(n)}$ and $\mathbf{w}^{(n)}$ are such that

$$D^{(n)} = \mathbf{u}^{(n)}\mathbf{w}^{(n)\mathrm{T}};$$

in Broyden's method, for example, (7.20) implies that

$$\mathbf{u}^{(n)} = \Delta^{(n)} + \sigma_n \mathbf{f}^{(n)}$$

and

$$\mathbf{w}^{(n)} = \frac{1}{\sigma_n \mathbf{p}^{(n)\mathrm{T}}\mathbf{p}^{(n)}}\,\mathbf{p}^{(n)}.$$

The problem is therefore to find $L^{(n+1)}$ and $Q^{(n+1)}$ such that

$$L^{(n+1)}Q^{(n+1)} = L^{(n)}Q^{(n)} + \mathbf{u}^{(n)}\mathbf{w}^{(n)\mathrm{T}}.$$

Since

$$Q^{(n)\mathrm{T}}Q^{(n)} = I,$$

$$\begin{aligned} J^{(n+1)} &= L^{(n)}Q^{(n)} + \mathbf{u}^{(n)}\mathbf{w}^{(n)\mathrm{T}}Q^{(n)\mathrm{T}}Q^{(n)} \\ &= (L^{(n)} + \mathbf{u}^{(n)}\mathbf{w}^{(n)\mathrm{T}}Q^{(n)\mathrm{T}})Q^{(n)}; \end{aligned}$$

write $\mathbf{z}^{(n)} \equiv Q^{(n)}\mathbf{w}^{(n)}$, then evaluate the orthogonal matrix Q such that

$$Q\mathbf{z}^{(n)} = \gamma_n \mathbf{e}_1 \equiv \begin{bmatrix} \gamma_n \\ 0 \\ \cdot \\ \cdot \\ \cdot \\ 0 \end{bmatrix}.$$

The matrix Q can be formed using *Givens' method* as described in Section 3.2.

Then, as $Q^{\mathrm{T}}Q = I$,

$$\begin{aligned} J^{(n+1)} &= (L^{(n)} + \mathbf{u}^{(n)}\mathbf{z}^{(n)\mathrm{T}})Q^{\mathrm{T}}QQ^{(n)} \\ &= (L^{(n)}Q^{\mathrm{T}} + \gamma_n \mathbf{u}^{(n)}\mathbf{e}_1^{\mathrm{T}})QQ^{(n)} \\ &= (G^{(n)} + \gamma_n \mathbf{u}^{(n)}\mathbf{e}_1^{\mathrm{T}})P^{(n)}. \end{aligned}$$

The matrices $G^{(n)}$ and $P^{(n)}$ can be determined simultaneously with Q and so this procedure is not as cumbersome as it might at first appear. That is, the augmented matrix $[\mathbf{z}^{(n)} \mid Q^{(n)} \mid L^{(n)\mathrm{T}}]$ is transformed into

$$Q[\mathbf{z}^{(n)} \mid Q^{(n)} \mid L^{(n)\mathrm{T}}] = [\gamma_n \mathbf{e}_1 \mid P^{(n)} \mid G^{(n)\mathrm{T}}]$$

without the need to determine Q explicitly. If the reduction is ordered correctly, then $G^{(n)}$ and

$$\bar{G}^{(n)} = G^{(n)} + \gamma_n \mathbf{u}^{(n)}\mathbf{e}_1^{\mathrm{T}}$$

are lower Hessenburg matrices; that is lower triangular plus a non-zero super-diagonal, i.e.

$$g_{ij} = 0, \qquad j - i > 1.$$

To complete the factorization of $J^{(n+1)}$, it only remains to construct, using Givens' transformations, an orthogonal matrix $\bar{Q}$ such that $\bar{G}^{(n)}\bar{Q}$ is lower triangul and then it follows that

$$\begin{aligned} J^{(n+1)} &= (\bar{G}^{(n)}\bar{Q})(\bar{Q}^{\mathrm{T}}P^{(n)}) \\ &= L^{(n+1)}Q^{(n+1)}. \end{aligned}$$

Again, the two transformations of $\bar{G}^{(n)}$ to $L^{(n+1)}$ and of $P^{(n)}$ to $Q^{(n+1)}$, can be performed simultaneously by computing $\bar{Q}^{\mathrm{T}}\ [P^{(n)} \mid \bar{G}^{(n)\mathrm{T}}]$. Thus it is not necessary to store either of the intermediate matrices Q and $\bar{Q}$ explicitly.

Numerical example Assume that

$$J^{(n)} = \frac{1}{3}\begin{bmatrix} -1 & 2 & 2 \\ 0 & 3 & 6 \\ 6 & 0 & 3 \end{bmatrix} \text{ then}$$

$$Q^{(n)} = \frac{1}{3}\begin{bmatrix} -1 & 2 & 2 \\ 2 & -1 & 2 \\ 2 & 2 & -1 \end{bmatrix} \text{ and } L^{(n)} = \begin{bmatrix} 1 & & \\ 2 & 1 & \\ 0 & 2 & 1 \end{bmatrix}. \text{ If } J^{(n+1)} = J^{(n)} + D^{(}$$

where $D^{(n)} \equiv \mathbf{u}\mathbf{w}^{\mathrm{T}} = \begin{bmatrix} 1 \\ 1 \\ -1 \end{bmatrix}[1 \quad -1 \quad 0]$ then $Q^{(n+1)}$ and $L^{(n+1)}$

can be found by modifying $Q^{(n)}$ and $L^{(n)}$ using Givens' transformations. It follows that

$$\mathbf{z}^{\mathrm{T}} = \mathbf{w}^{\mathrm{T}}Q^{(n)\mathrm{T}} = [-1 \quad 1 \quad 0]$$

and

$$[\mathbf{z} \mid Q^{(n)} \mid L^{(n)\mathrm{T}}] = \left[\begin{array}{c|ccc|ccc} -1 & -\frac{1}{3} & \frac{2}{3} & \frac{2}{3} & 1 & 2 & 0 \\ 1 & \frac{2}{3} & -\frac{1}{3} & \frac{2}{3} & 0 & 1 & 2 \\ 0 & \frac{2}{3} & \frac{2}{3} & -\frac{1}{3} & 0 & 0 & 1 \end{array}\right]$$

with

$$Q = \begin{bmatrix} -\frac{1}{\sqrt{2}} & \frac{1}{\sqrt{2}} & 0 \\ \frac{1}{\sqrt{2}} & -\frac{1}{\sqrt{2}} & 0 \\ 0 & 0 & 1 \end{bmatrix},$$

$$[\gamma \mathbf{e}_1 \,\vdots\, P^{(n)} \,\vdots\, G^{(n)\mathrm{T}}] = \left[\begin{array}{c|ccc|ccc} \sqrt{2} & \frac{1}{\sqrt{2}} & -\frac{1}{\sqrt{2}} & 0 & -\frac{1}{\sqrt{2}} & -\frac{1}{\sqrt{2}} & \frac{2}{\sqrt{2}} \\ 0 & \frac{1}{3\sqrt{2}} & \frac{1}{3\sqrt{2}} & \frac{4}{3\sqrt{2}} & \frac{1}{\sqrt{2}} & \frac{3}{\sqrt{2}} & \frac{2}{\sqrt{2}} \\ 0 & \frac{2}{3} & \frac{2}{3} & -\frac{1}{3} & 0 & 0 & 1 \end{array}\right].$$

Hence

$$\bar{G}^{(n)} = G^{(n)} + \gamma \mathbf{u}\, \mathbf{e}_1^{\mathrm{T}} = \frac{1}{\sqrt{2}} \begin{bmatrix} 1 & 1 & 0 \\ 1 & 3 & 0 \\ 0 & 2 & \sqrt{2} \end{bmatrix},$$

then with

$$\bar{Q} = \begin{bmatrix} \frac{1}{\sqrt{2}} & \frac{1}{\sqrt{2}} & 0 \\ -\frac{1}{\sqrt{2}} & \frac{1}{\sqrt{2}} & 0 \\ 0 & 0 & 1 \end{bmatrix}$$

$$[Q^{(n+1)} \,\vdots\, L^{(n+1)\mathrm{T}}] = \left[\begin{array}{ccc|ccc} \frac{2}{3} & -\frac{1}{3} & \frac{2}{3} & 1 & 2 & 1 \\ -\frac{1}{3} & \frac{2}{3} & \frac{2}{3} & 0 & 1 & 1 \\ \frac{2}{3} & \frac{2}{3} & -\frac{1}{3} & 0 & 0 & 1 \end{array}\right].$$

If the example is repeated using square-root-free Givens' transformations it is necessary to rewrite the factorization as

$$\begin{aligned} J^{(n+1)} &= (L^{(n)} + \mathbf{u}\mathbf{z}^{\mathrm{T}})Q^{(n)} \\ &= (\hat{L}^{(n)} + \mathbf{u}\hat{\mathbf{z}}^{\mathrm{T}})KQ^{\mathrm{T}}QK^{-1}Q^{(n)} \\ &= (G^{(n)} + \gamma \mathbf{u}\mathbf{e}_1^{\mathrm{T}})\bar{K}\bar{K}^{-1}P^{(n)} \\ &= \bar{G}^{(n)}\bar{K}\bar{K}^{-1}P^{(n)} \\ &= \bar{G}^{(n)}\bar{K}\bar{Q}\bar{Q}^{\mathrm{T}}\bar{K}^{-1}P^{(n)} \\ &= \bar{L}^{(n+1)}\bar{\bar{K}}\bar{\bar{K}}^{-1}\bar{Q}^{(n+1)} \\ &= L^{(n+1)}Q^{(n+1)}. \end{aligned}$$

The diagonal matrices K and $\bar{K}$ are not stored explicitly only K^2 and $\bar{K}^2$ are stored, the matrix $\bar{\bar{K}}^2$ is stored but $\bar{\bar{K}}$ must be computed at the final stage requiring a total of N square roots. The formulae for applying the Givens' transformations to the products such as $K^{-1}Q^{(n)}$ were derived in Exercise 3.3.

Working through the same example it follows that as initially $K = I$, then $\hat{L}^{(n)} = L^{(n)}$, etc. After the first transformation

$$\bar{K}^2 = \begin{bmatrix} \frac{1}{2} & & \\ & \frac{1}{2} & \\ & & 1 \end{bmatrix}$$

and

$$[\gamma \mathbf{e}_1 \,\vdots\, G^{(n)\mathrm{T}} \,\vdots\, P^{(n)}] = \left[\begin{array}{c|ccc|ccc} -2 & 1 & 1 & -2 & -\frac{1}{2} & \frac{1}{2} & 0 \\ 0 & 1 & 3 & 2 & \frac{1}{6} & \frac{1}{6} & \frac{2}{3} \\ 0 & 0 & 0 & 1 & \frac{2}{3} & \frac{2}{3} & -\frac{1}{3} \end{array}\right].$$

Then

$$\bar{G}^{(n)\mathrm{T}} = \begin{bmatrix} -1 & -1 & 0 \\ 1 & 3 & 2 \\ 0 & 0 & 1 \end{bmatrix}$$

and so after the second transformation

$$\bar{\bar{K}}^2 = \begin{bmatrix} \frac{1}{4} & & \\ & \frac{1}{4} & \\ & & 1 \end{bmatrix}$$

and

$$[\bar{L}^{(n+1)\mathrm{T}} \,\vdots\, \bar{Q}^{(n+1)}] = \left[\begin{array}{ccc|ccc} -2 & -4 & -2 & -\frac{1}{3} & \frac{1}{6} & -\frac{1}{3} \\ 0 & 2 & 2 & -\frac{1}{6} & \frac{1}{3} & \frac{1}{3} \\ 0 & 0 & 1 & \frac{2}{3} & \frac{2}{3} & -\frac{1}{3} \end{array}\right]$$

which after scaling of the first row by the negative square root and of the second row by the positive square root leads to the same result as before.

Exercises

7.7 Verify that a matrix of the form $[\mathbf{z} \,\vdots\, U]$, where U is $N \times N$ upper triangular, can be transformed into $[\mathbf{w} \,\vdots\, H]$, where $\mathbf{w} = \gamma \mathbf{e}_1$ and H is upper Hessenburg, using $N - 1$ Givens' transformations.

7.8. Verify that there are only $O(N^2)$ arithmetic operations involved in the modification of orthogonal factor using Givens' transformations.

(b) Using the structure of the Jacobian

The most frequently occurring 'structure' is *sparsity*; that is, very few of the components of the matrix are non-zero. Sometimes the non-zero components occur in a purely random way, but it is more likely that there is a precise and easily definable region of the matrix in which the non-zeros occur. For example, if the non-linear equations arise during the solution of non-linear differential equations by *finite-difference methods* or *finite-element methods* (cf. Sections 4.3 and 5.6) the Jacobian matrix may be banded or block banded (see Figures 18(a) and 18(b) respectively). Sparsity is a useful property to exploit if any Newton or modified Newton method is to be used.

In addition to the drawbacks that result from the presence of round-off errors, the implementation of quasi-Newton methods that update $H^{(n)}$ rather than $J^{(n)}$ can impose a greater restriction on the *size* of problem that can be solved on a particular machine. For if the Jacobian is sparse and structured, then only the non-zero components need be stored, whereas, even when $J^{(n)}$ is sparse, it is unlikely that $H^{(n)}$ will have a significant number of zero components and must therefore be stored in full.

It is also readily observed that, in general, a rank-one update will destroy the structure of the Jacobian; since for example if $J^{(n)}$ is banded it does not follow that

$$J^{(n+1)} = J^{(n)} + \mathbf{u}^{(n)}\mathbf{w}^{(n)\mathrm{T}}$$

will be similarly banded. Thus it is necessary to modify the formulation of a quasi-Newton method in order to produce a *structure-preserving algorithm.*

The method described below was first proposed by Schubert (1970) (see also Broyden, 1971), and it attempts to update the individual rows of $J^{(n)}$ using a different form of modification for each row, whilst still satisfying the quasi-Newton equation,

$$\sigma_n J^{(n+1)}\mathbf{p}^{(n)} = \Delta^{(n)}. \tag{7.21}$$

Those readers interested in the more general case, that allows $J^{(n)}$ to have some known *non-zero* constant components as well as known *zero* components, should consult the references cited above.

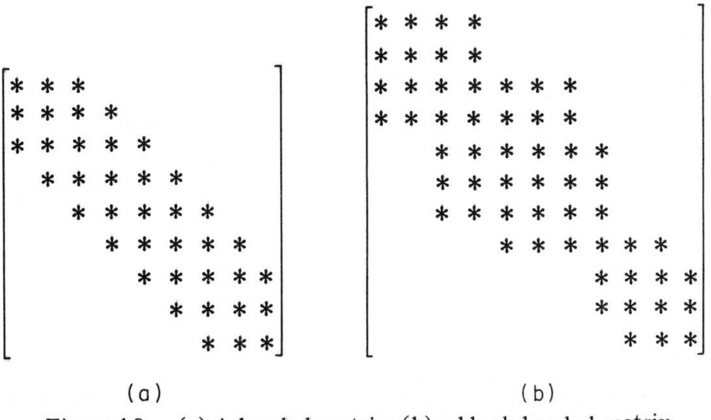

(a) (b)

Figure 18 (a) A banded matrix; (b) a block-banded matrix

Denote by

$$\mathbf{g}_i^{(k)\mathrm{T}} \equiv [g_{i1}^{(k)}, \ldots, g_{iN}^{(k)}]$$

the rows of $J^{(k)}$ $(k = n, n+1)$ and by $\Delta_i^{(n)}$ and $f_i^{(n)}$ the components of $\Delta^{(n)}$ and $\mathbf{f}^{(n)}$ respectively. Then (7.21) can be written as

$$\sigma_n \mathbf{g}_i^{(n+1)\mathrm{T}} \mathbf{p}^{(n)} = \Delta_i^{(n)}, \qquad i = 1, \ldots, N, \tag{7.22}$$

and the updating formula (7.20) as

$$\mathbf{g}_i^{(n+1)} = \mathbf{g}_i^{(n)} + \frac{\Delta_i^{(n)} + \sigma_n \mathbf{f}_i^{(n)}}{\sigma_n \mathbf{p}^{(n)\mathrm{T}} \mathbf{p}^{(n)}} \mathbf{p}^{(n)}, \qquad i = 1, \ldots, N. \tag{7.23}$$

The modification to preserve the structure is then as follows: for each i, to replace

$\mathbf{p}^{(n)} = [p_1^{(n)}, \ldots, p_N^{(n)}]^{\mathrm{T}}$ in (7.23) by

$$\mathbf{p}_i^{(n)} = [p_{i1}^{(n)}, \ldots, p_{iN}^{(n)}]^{\mathrm{T}},$$

where

$$p_{ij}^{(n)} = \begin{cases} p_j^{(n)}, & g_{ij}^{(n)} \neq 0 \\ 0, & g_{ij}^{(n)} = 0. \end{cases}$$

That is, $\mathbf{p}_i^{(n)}$ is a version of $\mathbf{p}^{(n)}$ that has been modified to have the desired pattern of non-zero components. It then follows from (7.23) that

$$\sigma_n \mathbf{g}_i^{(n+1)\mathrm{T}} \mathbf{p}^{(n)} = \sigma_n \mathbf{g}_i^{(n)\mathrm{T}} \mathbf{p}^{(n)} + \frac{\Delta_i^{(n)} + \sigma_n f_i^{(n)}}{\mathbf{p}_i^{(n)\mathrm{T}} \mathbf{p}_i^{(n)}} \mathbf{p}_i^{(n)\mathrm{T}} \mathbf{p}^{(n)}, \qquad i = 1, \ldots, N$$

hence as

$$\mathbf{g}_i^{(n)\mathrm{T}} \mathbf{p}^{(n)} = -f_i^{(n)} \qquad (\text{i.e., } J^{(n)} \mathbf{p}^{(n)} = -\mathbf{f}^{(n)}),$$

and

$$\mathbf{p}_i^{(n)\mathrm{T}} \mathbf{p}^{(n)} = \mathbf{p}_i^{(n)\mathrm{T}} \mathbf{p}_i^{(n)}$$

it follows that (7.22) is satisfied.

In addition to preserving sparsity and symmetry it is also possible to define update formulae that preserve positive definiteness in the matrices $J^{(n)}$, but that is beyond the scope of this book.

Exercise

7.9. Verify that if $J^{(n)}$ is a diagonal matrix, that is,

$$\mathbf{f}(\mathbf{x}) = [f_1(x_1), \ldots, f_N(x_N)]^{\mathrm{T}}$$

then (7.23) is equivalent to applying the *single-variable* secant method simultaneously to the functions f_i $(i = 1, \ldots, N)$, provided $\sigma_n = 1$.

7.4 Gradient Methods

One of the most significant drawbacks of quasi-Newton methods is a failure to converge owing to the Jacobian matrix being singular. Usually at – or very close to – the solution (Bard, 1968; Powell, 1970).

An alternative approach is to reformulate the problem as the *minimization of a sum of squares – a non-linear least-squares problem*. The linear least-squares problem is described in Section 3.3. Let

$$F(\mathbf{x}) = \mathbf{f}(\mathbf{x})^{\mathrm{T}}\mathbf{f}(\mathbf{x}); \tag{7.24}$$

then $F(\mathbf{x})$ is a non-negative function such that the zero of $\mathbf{f}(\mathbf{x})$ gives a *global minimum* of $F(\mathbf{x})$. At any point $\mathbf{x}$, the direction in which $F(\mathbf{x})$ decreases most rapidly is $-\nabla F(\mathbf{x})$; thus provided $\nabla F(\mathbf{x}) \not\equiv 0$, it is possible to find $\sigma > 0$ such that

$$F(\mathbf{x} - \sigma\nabla F(\mathbf{x})) < F(\mathbf{x}).$$

When the value of σ is chosen to minimize $F(\mathbf{x} - \sigma\nabla F(x))$, the procedure is known as the *method of steepest descent*.* Since $F(\mathbf{x})$ is of the particular form (7.24), the gradient vector $\nabla F(\mathbf{x})$ is given by

$$\nabla F(\mathbf{x}) = 2\{J(\mathbf{x})\}^{\mathrm{T}}\mathbf{f}(\mathbf{x}),$$

where J is the Jacobian matrix of $\mathbf{f}$. Note that this method may converge to a *local minimum* at which

$$\nabla F(\mathbf{x}) = 0$$

but

$$\mathbf{f}(\mathbf{x}) \neq 0,$$

that is to a point at which $J(\mathbf{x})$ is singular. While it can be shown that descent methods of the form

$$\mathbf{x}^{(n+1)} = \mathbf{x}^{(n)} - \sigma_n\nabla F(\mathbf{x}^{(n)}), \qquad n = 0, 1, \ldots,$$

do converge – at least to a local minimum (Goldstein, 1967, pp. 28–34) – the convergence tends to be very slow as the minimum is approached.

If $J_\nabla(\mathbf{x})$ is the Jacobian of $\nabla F(\mathbf{x})$, it can be written as

$$J_\nabla(\mathbf{x}) \equiv 2\{J(\mathbf{x})\}^{\mathrm{T}}J(\mathbf{x}) + 2\sum_{i=1}^{N} H_i(\mathbf{x})f_i(\mathbf{x}), \tag{7.25}$$

where $H_i(\mathbf{x})$ is the matrix of second derivatives – the *Hessian matrix* – of the ith component of $\mathbf{f}(\mathbf{x})$ ($i = 1, \ldots, N$). The Newton method applied to the system $\nabla F(\mathbf{x}) = 0$ leads to a correction

$$\boldsymbol{\delta}^{(n)} \equiv \mathbf{x}^{(n+1)} - \mathbf{x}^{(n)},$$

*The history of this method can be traced back to a paper by Cauchy in 1847.

that satisfies

$$J_\nabla(\mathbf{x}^{(n)})\boldsymbol{\delta}^{(n)} = -\nabla F(\mathbf{x}^{(n)}).$$

If, however, the second-derivative terms in (7.25) are ignored, the method reduces to the so-called *Gauss–Newton method*, in which

$$\{J(\mathbf{x}^{(n)})\}^{\mathrm{T}} J(\mathbf{x}^{(n)})\boldsymbol{\delta}^{(n)} = -\{J(\mathbf{x}^{(n)})\}^{\mathrm{T}} \mathbf{f}(\mathbf{x}^{(n)}). \tag{7.26}$$

This method has deficiencies similar to the original Newton method, namely that although convergence – when it occurs – is rapid, the method frequently diverges and also breaks down when $J(\mathbf{x}^{(n)})$ becomes singular. In the *modified Gauss–Newton* method, the correction is

$$\boldsymbol{\delta}^{(n)} = \sigma_n \mathbf{p}^{(n)},$$

where $\mathbf{p}^{(n)}$ replaces $\boldsymbol{\delta}^{(n)}$ in (7.26) and the parameter σ_n is chosen such that $F(\mathbf{x}^{(n+1)}) < F(\mathbf{x}^{(n)})$; again it is not necessary to minimize $F(\mathbf{x}^{(n)} + \sigma_n \mathbf{p}^{(n)})$ with respect to σ_n.

A way of overcoming the deficiencies of both the steepest descent and Gauss–Newton methods is to combine them into a method that has the best features of both. One such procedure is the *Levenberg–Marquardt method.* The correction at any step is then given by

$$(\{J(\mathbf{x}^{(n)})\}^{\mathrm{T}} J(\mathbf{x}^{(n)}) + \lambda I)\boldsymbol{\delta}^{(n)}(\lambda) = -\{J(\mathbf{x}^{(n)})\}^{\mathrm{T}} \mathbf{f}(\mathbf{x}^{(n)}), \qquad n = 0, 1, \ldots . \tag{7.27}$$

The parameter λ is first set to zero, thus initially the method computes a Gauss–Newton step. The algorithm is then as follows: given λ, compute $\boldsymbol{\delta}^{(n)}(\lambda)$ from (7.27), and then if

$$F(\mathbf{x}^{(n)} + \boldsymbol{\delta}^{(n)}(\lambda)) \begin{cases} \leqslant F(\mathbf{x}^{(n)}), \text{ accept } \mathbf{x}^{(n+1)} = \mathbf{x}^{(n)} + \boldsymbol{\delta}^{(n)}(\lambda) \\ > F(\mathbf{x}^{(n)}), \text{ increase } \lambda \text{ and recompute } \boldsymbol{\delta}^{(n)}(\lambda). \end{cases}$$

Note that as λ increases, it follows from (7.27) that the direction of $\boldsymbol{\delta}^{(n)}(\lambda)$ tends to the steepest descent direction, while the length of $\boldsymbol{\delta}^{(n)}(\lambda)$ tends to zero. Thus the test

$$F(\mathbf{x}^{(n)} + \boldsymbol{\delta}^{(n)}(\lambda)) \leqslant F(\mathbf{x}^{(n)}),$$

will always be satisfied if λ is sufficiently large. A good account of the computational problems involved in implementing this technique are given by Lill (1976).

An alternative procedure for combining the steepest descent and Newton methods has been suggested by Powell (1970): at each step both the Newton correction

$$\mathbf{p}^{(n)} = -\{J(\mathbf{x}^{(n)})\}^{-1}\mathbf{f}(\mathbf{x}^{(n)}) \tag{7.28}$$

and the steepest-descent correction

$$\mathbf{q}^{(n)} = -2\{J(\mathbf{x}^{(n)})\}^{\mathrm{T}}\mathbf{f}(x^{(n)}) \tag{7.29}$$

are computed. Then the step taken is a linear combination

$$\boldsymbol{\delta}^{(n)} = \alpha_n \mathbf{p}^{(n)} + \beta_n \mathbf{q}^{(n)}.$$

This approach has advantages when the Jacobian in (7.29) and its inverse in (7.28) are replaced by approximations that are updated at each iteration in a manner analogous to the quasi-Newton methods. The modifications used can be found in the reference above. Discussions of least-squares methods can also be found in Kowalik and Osborne (1968).

8

Other Methods

8.1 Multiple Roots

A numerical example Newton's method applied to the equation $x \sin x = 0$ with $x^{(0)} = 2.5$ and $x^{(0)} = 0.5$ lead to the sequences given in Table 11.

Table 11

n	x	x
0	2.5	0.5
1	3.5654	0.2389
2	3.1654	0.1183
3	3.1418	0.0590
4	3.1416	0.0295
5		0.0247
6		0.0074
7		0.0037
8		0.0018
9		0.0009
10		0.0005
11		0.0002
12		0.0001
13		0.0000

The reason for the slow convergence when $x^{(0)} = 0.5$ is that the sequence $\{x^{(n)}\}$ tends to zero, at which point there is a double root ($f(0) = f'(0) = 0; f''(0) \neq 0$). Note that the sequence is such that

$$\frac{x^{(n+1)}}{x^{(n)}} \approx \tfrac{1}{2}, \qquad n = 0, 1, \ldots .$$

The quadratic convergence of Newton's method was proved in Chapter 5, subject to the provision that $f'(x) \neq 0$ at the root, and in the above example the convergence to zero was in fact linear. All the orders of convergence given in Chapters 5 and 6 assume that $f'(x) \neq 0$ at the root and thus they are not valid for multiple roots.

Lemma *Newton's method converges linearly to a multiple root.* Δ

Proof Assume that

$$f(x) \equiv (x - \alpha)^m g(x), \tag{8.1}$$

where

$$g(\alpha) \neq 0.$$

Then

$$f'(x) = (x - \alpha)^m g'(x) + m(x - \alpha)^{m-1} g(x).$$

If the sequence $\{x^{(n)}\}$ is defined by Newton's method it follows that

$$\begin{aligned} x^{(n+1)} - \alpha &= x^{(n)} - \alpha - \frac{f(x^{(n)})}{f'(x^{(n)})} \\ &= \frac{(m-1)g(x^{(n)}) + (x^{(n)} - \alpha)g'(x)}{mg(x^{(n)}) + (x^{(n)} - \alpha)g'(x)}(x^{(n)} - \alpha) \end{aligned} \tag{8.2}$$

and

$$\lim_{n\to\infty} \frac{x^{(n+1)} - \alpha}{x^{(n)} - \alpha} = \frac{m-1}{m}. \tag{8.3}$$

Thus the convergence is linear, and if the multiplicity is m, the asymptotic error constant is $(m - 1)/m$; hence the factor $\frac{1}{2}$ in the above example. Δ

The best-known method for multiple roots is due to Schröder (1870):

$$x^{(n+1)} = x^{(n)} - m\frac{f(x^{(n)})}{f'(x^{(n)})}, \qquad n = 0, 1, \ldots,$$

where m is the multiplicity of the root.

Lemma *Schröder's method is quadratically convergent.* Δ

Proof When $f(x)$ is given by (8.1) it follows by analogy with (8.2) that Schröder's method leads to

$$\begin{aligned} x^{(n+1)} - \alpha &= x^{(n)} - \alpha - m\frac{f(x^{(n)})}{f'(x^{(n)})} \\ &= (x^{(n)} - \alpha)^2 \frac{g'(x^{(n)})}{mg(x^{(n)}) + (x^{(n)} - \alpha)g'(x^{(n)})}, \end{aligned}$$

and thus

$$\lim_{n\to\infty} \frac{x^{(n+1)} - \alpha}{(x^{(n)} - \alpha)^2} = \frac{g'(\alpha)}{mg(\alpha)}. \quad \Delta \tag{8.4}$$

Note that

$$f^{(m)}(\alpha) = m!g(\alpha)$$

and

$$f^{(m+1)}(\alpha) = (m+1)!g'(\alpha),$$

so that the asymptotic error constant in (8.4) can be written as

$$\frac{g'(\alpha)}{mg(\alpha)} = \frac{f^{(m+1)}(\alpha)}{m(m+1)f^{(m)}(\alpha)}.$$

One formulation of Schröder's method that leads directly to a proof of the quadratic convergence is to apply Newton's method to

$$h(x) \equiv \{f(x)\}^{1/m}.$$

This is a function with a simple root at $x = \alpha$, and hence the quadratic convergence of Newton's method is valid.

In order to make use of Schröder's method, it is necessary to know the multiplicity. Hence as the multiplicity is unknown in general, it is necessary to have a procedure for estimating it.

(1) If the sequence $\{x^{(n)}\}$ is generated by Newton's method, then it follows from (8.2) (see Exercise 8.3) that when $m > 1$,

$$\lim_{n\to\infty} \frac{x^{(n)} - x^{(n+1)}}{x^{(n)} - 2x^{(n+1)} + x^{(n+2)}} = m. \tag{8.5}$$

(2) For any sequence $\{x_n\}$ which has the root α as the limit,

$$\lim_{n\to\infty} \frac{\ln|f(x)|}{\ln\left|\frac{f(x)}{f'(x)}\right|} = m. \tag{8.6}$$

The use of this second method of estimating the multiplicity has been considered in detail by Traub (1964).

Exercises

8.1. Verify the Newton's method applied to $h(x) \equiv \{f(x)\}^{1/m}$ leads to Schröder's method.

8.2. Verify that Chebychev's method (5.18) leads to the third-order method

$$x^{(n+1)} = x^{(n)} - \frac{m}{2}(3-m)\frac{f(x)}{f'(x)} - m^2\frac{f''(x^{(n)})f(x^{(n)})^2}{f'(x^{(n)})^3}, \qquad n = 0, 1, \ldots,$$

when applied to $h(x) \equiv \{f(x)\}^{1/m}$.

8.3. Prove that if

$$x^{(n+1)} = x^{(n)} - \frac{f(x^{(n)})}{f'(x^{(n)})}, \qquad n = 0, 1, \ldots,$$

then for a root of multiplicity m (>1),

$$\lim_{n\to\infty} \frac{x^{(n)} - x^{(n+1)}}{x^{(n-1)} - x^{(n)}} = \frac{m-1}{m}$$

and hence verify (8.5).

8.2 Aitken Δ^2-acceleration

When an iteration

$$x^{(n+1)} = \varphi(x^{(n)}), \qquad n = 0, 1, \ldots,$$

leads to a linearly convergent sequence $\{x^{(n)}\}$, one method of improving the convergence is to use *extrapolation* in the form of Aitken Δ^2-acceleration (Aitken, 1926).

Let the sequence $\{x^{(n)}\}$ be defined by

$$\bar{x}^{(n)} = x^{(n+2)} - \frac{(x^{(n+2)} - x^{(n+1)})^2}{x^{(n)} - 2x^{(n+1)} + x^{(n+2)}}, \qquad n = 0, 1, \ldots. \tag{8.7}$$

Equivalently this formula for accelerating the convergence can be written as

$$\bar{x}^{(n)} = x^{(n)} - \frac{(x^{(n+1)} - x^{(n)})^2}{x^{(n)} - 2x^{(n+1)} + x^{(n+2)}}, \qquad n = 0, 1, \ldots,$$

or in terms of the *forward difference operator* Δ, as

$$\bar{x}^{(n)} = x^{(n)} - \frac{(\Delta x^{(n)})^2}{\Delta^2 x^{(n)}}, \qquad n = 0, 1, \ldots,$$

and hence the name Δ^2-acceleration.

Lemma *If the sequence* $\{x^{(n)}\}$ *converges linearly, the sequence* $\{\bar{x}^{(n)}\}$ *converges* linearly *and ultimately monotonically, and is such that*

$$\lim_{n\to\infty} \frac{\bar{x}^{(n)} - \alpha}{(x^{(n)} - \alpha)^2} = \frac{1}{2}\,\frac{\varphi'(\alpha)\varphi''(\alpha)}{\varphi'(\alpha) - 1}. \qquad \Delta$$

Proof (Traub, 1964, pp. 265–7). Let

$$e_n \equiv x^{(n)} - \alpha, \qquad n = 0, 1, \ldots, \text{and } \gamma \equiv \varphi'(\alpha);$$

then

$$e_{n+1} = (\gamma + \sigma_n)e_n, \qquad n = 0, 1, \ldots, \tag{8.8}$$

where

$$\sigma_n = e_n \frac{\varphi''(\xi_n)}{2}, \qquad n = 0, 1, \ldots, \tag{8.8a}$$

for some ξ_n between $x^{(n)}$ and α. It then follows that

$$e_{n+2} = (\gamma + \sigma_n)(\gamma + \sigma_{n+1})e_n, \tag{8.9}$$

and if $\bar{e}_n \equiv \bar{x}^{(n)} - \alpha$ it follows from (8.7) that

$$\bar{e}_n = e_{n+2} - \frac{(e_{n+2} - e_{n+1})^2}{e_n - 2e_{n+1} + e_{n+2}}$$

and so

$$\frac{\bar{e}_n}{e_n^2} = \frac{\gamma + \sigma_n}{(\gamma - 1)^2 + \lambda_n} \left\{ \frac{\sigma_{n+1} - \sigma_n}{e_n} \right\}, \tag{8.10}$$

where

$$\lambda_n \equiv \gamma(\sigma_n + \sigma_{n+1}) - 2\sigma_n + \sigma_n\sigma_{n+1}.$$

It follows from (8.8) and (8.8a) that

$$\lim_{n\to\infty} \frac{\sigma_{n+1} - \sigma_n}{e_n} = (\varphi'(\alpha) - 1)\frac{\varphi''(\alpha)}{2},$$

and so it follows that

$$\lim_{n\to\infty} \frac{\bar{e}_n}{e_n^2} = \frac{\varphi'(\alpha)}{\varphi'(\alpha) - 1} \frac{\varphi''(\alpha)}{2}, \tag{8.11}$$

which is the desired result. The linear and hence monotone convergence follows directly from (8.8) since

$$\lim_{n\to\infty} \frac{e_{n+1}}{e_n} = \varphi'(\alpha),$$

and so

$$\lim_{n\to\infty} \frac{\bar{e}_{n+1}}{\bar{e}_n} = \{\varphi'(\alpha)\}^2. \quad \Delta$$

Steffensen iteration

The Δ^2-process can be reformulated to provide a quadratically convergent method known as *Steffensen iteration* (Steffensen, 1933). The iteration defines a sequence $\{x^{(n)}\}$ such that given the function $\varphi(x)$

$$x_0^{(n)} = x^{(n)},$$

$$x_1^{(n)} = \varphi(x_0^{(n)}),$$

$$x_2^{(n)} = \varphi(x_1^{(n)}), \qquad n = 0, 1, \ldots,$$

then

$$x^{(n+1)} = x^{(n)} - \frac{(x_0^{(n)} - x_1^{(n)})^2}{x_0^{(n)} - 2x_1^{(n)} + x_2^{(n)}}. \tag{8.12}$$

It follows by analogy with (8.7) that the sequence is quadratically convergent and that the asymptotic error constant is

$$\frac{\varphi'(\alpha)}{\varphi'(\alpha) - 1} \frac{\varphi''(\alpha)}{2}.$$

If a sequence $\{x^{(n)}\}$ derived from the iteration $x^{(n+1)} = \varphi(x^{(n)})$ $(n = 0, 1, \ldots)$ diverges, then it follows that the sequence $\{\bar{x}^{(n)}\}$ derived from the Δ^2-procedure also diverges, whereas if $x^{(0)} - \alpha$ is sufficiently small, the quadratically convergent Steffensen iteration (8.12) will still converge (see Exercise 8.7).

Exercises

8.4. By considering the iteration

$$\begin{bmatrix} x^{(n+1)} \\ y^{(n+1)} \end{bmatrix} = \begin{bmatrix} \alpha & \beta \\ \gamma & \delta \end{bmatrix} \begin{bmatrix} x^{(n)} \\ y^{(n)} \end{bmatrix} + \begin{bmatrix} b \\ c \end{bmatrix}, \qquad n = 0, 1, \ldots,$$

verify that it is not possible to accelerate the convergence of individual components of an iteration in more than one variable unless the error equations (8.8) and (8.8a) are valid for each component. (*Hint*: the Δ^2-process is quadratically convergent and therefore obtains the root of a linear equation in one step, i.e., $\bar{e}_1 = 0$. Is this true for the linear system above when $\beta \neq 0$, for example?)

8.5. Given that ∇ is the *backward difference operator*, verify that the following are equivalent formulations, assuming that exact arithmetic is used,

$$\bar{x}^{(n)} = x^{(n+2)} - \frac{(\nabla x^{(n+2)})^2}{\nabla^2 x^{(n+2)}},$$

$$\bar{x}^{(n)} = x^{(n)} - \frac{(\Delta x^{(n)})^2}{\Delta^2 x^{(n)}}$$

and

$$\bar{x}^{(n)} = \frac{\det\begin{bmatrix} x^{(n)} & x^{(n+1)} \\ \Delta x^{(n)} & \Delta x^{(n+1)} \end{bmatrix}}{\det\begin{bmatrix} 1 & 1 \\ \Delta x^{(n)} & \Delta x^{(n+1)} \end{bmatrix}}. \tag{8.13}$$

(Formula (8.13) leads to a formulation of the Δ^2-acceleration, from which can be derived higher-order extrapolation methods such as the *ϵ-algorithm* of Wynn. A detailed discussion of such developments is to be found in Householder (1970) or Henrici (1964).)

8.6. Verify numerically that when

$$\varphi(x) = \frac{1}{7}(x^3 - 6),$$

the Steffensen iteration converges when $x^{(0)} = 2.2$, whereas the basic iteration

$$x^{(n+1)} = \varphi(x^{(n)}), \qquad n = 0, 1, \ldots,$$

does not converge (see Table 1).

8.7. Prove that if the initial approximation $x^{(0)}$ is sufficiently close to the solution, the Steffensen iteration converges quadratically whether or not the iteration

$$x^{(n+1)} = \varphi(x^{(n)}), \qquad n = 0, 1, \ldots,$$

converges from the same initial approximation.

*8.3 Multi-stage Methods

In Section 5.7, a modification of Newton's method is proposed, whereby the same Jacobian matrix is retained for a number of successive steps and not recomputed as $J(x^{(n)})$ every time. If it is assumed that the Jacobian is retained for k steps, it is possible to formulate k steps of such a method as one iteration of a *k-stage method*. One can then compare the efficiency of such an approach with the various alternatives (see Chapter 9 for such a comparison). In order to simplify the presentation the method will be applied to a single equation only.

The sequence $\{x^{(n)}\}$ is derived from the iteration

$$\begin{aligned} x_0^{(n)} &= x^{(n)}, \\ x_{j+1}^{(n)} &= x_j^{(n)} - \frac{f(x_j^{(n)})}{f'(x_0^{(n)})}, \qquad j = 0, 1, \ldots, k-1, \end{aligned} \tag{8.14}$$

*Can be omitted on first reading.

then

$$x^{(n+1)} = x_k^{(n)}.$$

Clearly the above iteration is equivalent to a modification of Newton's method in which the derivative is recomputed after every k steps. Another example of a k stage method ($k = 2$) is the Steffensen iteration defined by (8.12).

Lemma *The order of the k-stage method defined by (8.14) is $k + 1$ and the asymptotic error constant is*

$$\frac{1}{2}\left\{\frac{f''(\alpha)}{f'(\alpha)}\right\}^k$$

when $f'(\alpha) \neq 0$. △

Proof The proof is inductive and it will show that if $f'(\alpha) \neq 0$, then

$$\lim_{n\to\infty} \frac{x_j^{(n)} - \alpha}{(x^{(n)} - \alpha)^{j+1}} = \frac{1}{2}\left\{\frac{f''(\alpha)}{f'(\alpha)}\right\}^j, \qquad (j = 1, 2, \ldots, k). \tag{8.15}$$

This result is obvious for $j = 1$, since it is simply a statement of the quadratic convergence of Newton's method (see Section 5.6). Thus in order to complete the proof it is necessary and sufficient to prove that if (8.15) is valid for any j $(1 \leqslant j < k)$, it is also valid when j is replaced by $j + 1$.

From (8.14),

$$x_{j+1}^{(n)} - \alpha = x_j^{(n)} - \alpha - \frac{f(x_j^{(n)})}{f'(x^{(n)})},$$

and as $f(\alpha) = 0$, it follows from Taylor's theorem that there exists $\xi_j^{(n)}$, between $x_j^{(n)}$ and α, such that

$$f(x_j^{(n)}) = f'(\alpha)(x_j^{(n)} - \alpha) + \frac{f''(\xi_j^{(n)})}{2}(x_j^{(n)} - \alpha)^2.$$

Thus

$$x_{j+1}^{(n)} - \alpha = \frac{(x_j^{(n)} - \alpha)}{f'(x^{(n)})}\left\{f'(x^{(n)}) - f'(\alpha) - \frac{f''(\xi_j^{(n)})}{2}(x_j^{(n)} - \alpha)\right\},$$

and from the mean value theorem it follows that there exists $\eta^{(n)}$, between $x^{(n)}$ and α, such that

$$\frac{x_{j+1}^{(n)} - \alpha}{x_j^{(n)} - \alpha} = \frac{1}{f'(x^{(n)})}\left\{(x^{(n)} - \alpha)f''(\eta^{(n)}) - \frac{f''(\xi_j^{(n)})}{2}(x_j^{(n)} - \alpha)\right\}.$$

As the iteration converges, it follows that both $\eta^{(n)}$ and $\xi_j^{(n)}$ tend to α, and from (8.15) it follows that

$$\lim_{n\to\infty} \frac{x_j^{(n)} - \alpha}{x^{(n)} - \alpha} = 0.$$

Hence

$$\lim_{n\to\infty} \frac{x_{j+1}^{(n)} - \alpha}{(x_j^{(n)} - \alpha)(x^{(n)} - \alpha)} = \frac{f''(\alpha)}{f'(\alpha)}, \tag{8.16}$$

which can be combined with (8.15) to give the desired result. Δ

Note that (8.16) shows that as k is increased, the convergence of the sequence $x_j^{(n)}$ $(j = 1, 2, \ldots, k)$ tends to become linear, with an asymptotic error constant

$$\frac{f''(\alpha)}{f'(\alpha)} (x^{(n)} - \alpha).$$

Thus the sequence should not be continued for too long before the derivative is recomputed.

An alternative approach has been suggested for problems in which the derivative is significantly *easier* to compute than the function itself. One example of the type of method produced is a two-stage method of the form

$$w_1^{(n)} = \frac{f(x^{(n)})}{f'(x^{(n)})},$$

$$w_2^{(n)} = \frac{f(x^{(n)})}{f'(x^{(n)}) + \omega\, w_1^{(n)}}, \qquad n = 0, 1, \ldots,$$

then

$$x^{(n+1)} = x^{(n)} + a_1 w_1^{(n)} + a_2 w_2^{(n)}, \tag{8.17}$$

where the parameters ω, a_1 and a_2 are chosen to maximize the order (Traub, 1964; Jarratt, 1970). This type of k-stage formula can be compared with k-stage Runge–Kutta methods for the numerical solution of ordinary differential equations (see Lambert, 1973).

Exercises

8.8. Verify that it is possible to apply the Aitken Δ^2-formula to three successive values $x_j^{(n)}, x_{j+1}^{(n)}$ and $x_{j+2}^{(n)}$ $(j \geq 1)$ in (8.14) and derive an approximation that is of order $j + 4$; that is Aitken's Δ^2-formula can be used to accelerate superlinear convergence.

.9. Using Taylor's theorem verify that $\omega = -1, a_1 = \frac{1}{2}, a_2 = \frac{1}{2}$ leads to a third-order method.

*8.4 A Continuation Method

The basic method (Davidenko, 1953) is to define a one-parameter family of unctions $\mathbf{f}^{(\theta)}(\mathbf{x})$ such that the solution of

$$\mathbf{f}^{(0)}(\mathbf{x}) = 0$$

known and such that

$$\mathbf{f}^{(1)}(\mathbf{x}) \equiv \mathbf{f}(\mathbf{x}).$$

Clearly one possible definition is

$$\mathbf{f}^{(\theta)}(\mathbf{x}) \equiv \mathbf{f}(\mathbf{x}) - \mathbf{f}(\mathbf{x}^{(0)})(1-\theta) \qquad (8.18)$$

or an *arbitrary* value of $\mathbf{x}^{(0)}$.

Then $\mathbf{x}(\theta)$ $(0 \leqslant \theta \leqslant 1)$ – called the *Davidenko path* – is a continuous function f θ such that

$$\mathbf{f}^{(\theta)}(\mathbf{x}(\theta)) = 0, \qquad 0 \leqslant \theta \leqslant 1 \qquad (8.19)$$

nd the desired solution is $\mathbf{x}(1)$. The function $\mathbf{x}(\theta)$ can be computed in various vays:

(1) For some $\Theta \in (0, 1)$ compute the sequence $\mathbf{x}(\theta^{(k)})$, where $\theta^{(k)} \equiv 1 - \Theta^k$ $k = 0, 1, \ldots$); that is, given $\mathbf{x}(\theta^{(k-1)})$, use an iterative method such as Newton's nethod to obtain the solution of (8.19) with $\theta = \theta^{(k)}$ using $\mathbf{x}(\theta^{(k-1)})$ as a first pproximation. Note that as only $\mathbf{x}(1)$ is required, it is *not necessary to solve each of the intermediate problems very accurately*. Then, when Θ^k is sufficiently small, ake $\mathbf{x}(\theta^{(k)})$ as an approximation to the solution of

$$\mathbf{f}(\mathbf{x}) = 0.$$

This approximation could be accepted as sufficiently accurate or it could be used s a first approximation in an iteration.

An alternative strategy is to find the position of the Davidenko path at $\theta = k/K$ $(k = 0, 1, \ldots, K)$ for some K rather than at $\theta = 1 - \Theta^k$.

(2) From (8.18) and (8.19) it follows that

$$\mathbf{f}(\mathbf{x}(\theta)) = \mathbf{f}(\mathbf{x}^{(0)})(1-\theta);$$

thus, differentiating with respect to θ leads to the system of equations

$$J(\theta)\dot{\mathbf{x}}(\theta) = -\mathbf{f}(\mathbf{x}^{(0)}), \qquad 0 \leqslant \theta \leqslant 1, \qquad (8.20)$$

*Can be omitted on first reading.

where J is the Jacobian matrix $\{\partial f_i/\partial x_j\}$ evaluated at $\mathbf{x} = \mathbf{x}(\theta)$ and

$$\dot{\mathbf{x}} = \left[\frac{dx_1}{d\theta}, \ldots, \frac{dx_N}{d\theta}\right]^T,$$

subject to the initial condition

$$\mathbf{x}(0) = \mathbf{x}^{(0)}.$$

The sytem of equations would then be integrated from $\theta = 0$ to $\theta = 1$ using a standard numerical technique such as a predictor–corrector method (see Lambert, 1973). The Davidenko method, with any of these strategies for determining the Davidenko path, is primarily a method for solving an equation (or system) when no reasonable starting approximation exists.

9

Comparison of Methods

Comparing different methods for solving algebraic equations involves at least three different criteria:

(1) The number of iterations required to attain a desired accuracy: It follows from the earlier chapters that, if the initial approximation is sufficiently accurate, then this is directly related to the order of the iteration and the asymptotic error constant.

(2) The amount of work necessary to complete each individual iteration; it is invariably assumed that the majority of the calculation is involved in evaluating the function $f(x)$ and if necessary its derivatives.

(3) The robustness of the method; it is necessary to know in advance whether or not the method will in fact converge to the desired solution.

The first two criteria have been combined to provide an *index of efficiency* for the methods. If p is the order of a method and d, called the *informational usage*, is the number of function and/or derivative evaluations necessary at each step, one possible index that has been suggested is $I_0 \equiv p/d$. This is simple to compute, but it has one drawback – the value of the index is dependent upon the number of steps of the iteration considered. If the order of one step of a method is p, the order of two steps is p^2; if the informational usage in one step is d, then the informational usage in two steps is $2d$. Thus if the index of one step is p/d the index of two steps is $p^2/2d$, which is only the same for $p = 2$.

An alternative index that can be applied to any number of steps and lead to the same result is $I_1 \equiv p^{1/d}$, as the corresponding index for two steps would be

$$(p^2)^{1/2d} = p^{1/d}.$$

This form is usually preferred as it is related to the overall cost of finding the root, where cost is assumed to be proportional to computer time. A significant defect of both these indices is that they take account of neither the asymptotic error constant, nor the difference in cost of function and derivative evaluations, and are very crude measures indeed.

9.1 Comparison of Newton and Secant Iterations

One significant assumption has to be made before a comparison of methods is feasible; that is that the iterates satisfy

$$(x^{(n+1)} - \alpha) = (x^{(n)} - \alpha)^p C \tag{9.1}$$

for all n and not just in the limit as n tends to infinity. Let

$$k_n \equiv -\log_{10} | x^{(n)} - \alpha |, \qquad n = 0, 1, \ldots;$$

then k_n is (approximately) the number of decimal places in $x^{(n)}$ that are correct. It follows from (9.1) that

$$k_{n+1} = pk_n - \log_{10} | C |, \qquad n = 0, 1, \ldots,$$

which has the general solution

$$k_n = p^n k_0 + \left(\frac{p^n - 1}{1 - p}\right) \log_{10} | C |, \qquad n = 0, 1, \ldots, \tag{9.2}$$

(see, for example, Levy and Lessman (1959), Chapter 4). If it is possible to estimate the numerical values of k_1 and C, it is possible to use (9.2) to estimate the number of iterations required to achieve a given accuracy.

As an example Newton's method ($p = 2$, $C = f''(\alpha)/2f'(\alpha)$) is applied to the function

$$f(x) \equiv x^2 - (1 - x)^5$$

to find a root in the interval $[0, 1]$. Assume that $x_0 = 0.5$, that

$$C \approx \frac{f''(0.5)}{2f'(0.5)} = -\frac{4}{21}$$

and that

$$| x^{(0)} - \alpha | \approx 0.5;$$

hence

$$k_0 \approx -\log_{10} \tfrac{1}{2} = 0.3010.$$

Then

$$\begin{aligned} k_n &= 2^n 0.3010 + (2^n - 1)\, 0.7201 \\ &= 2^n 1.0211 - 0.7201 \end{aligned} \tag{9.3}$$

and if q decimal places are required it follows that

$$| x^{(n)} - \alpha | < 0.5\,\mathrm{E} - q$$

and hence

$$k_n \geqslant 0.3010 + q.$$

›mbining the results it follows that

$$(2^n - 1)1.0211 > q.$$

able 12 Newton's method applied to x^2
$(1-x)^5 = 0$

k_n	Estimate	Actual
0.3010	0.5	0.145
1.3221	0.047	0.0126
3.3644	4.3 E − 4	1.8 E − 4
7.4487	3.6 E − 8	3.8 E − 8
15.6175	2.4 E − 16	1.6 E − 15
31.9551	1.1 E − 32	6.3 E − 29

Given two methods, with orders p_1 and p_2 and symptotic error constants C_1 ıd C_2 respectively, it is possible to use (9.2) to compare the number of iterations ›quired to achieve the same order, assuming the iterations use the same initial ›proximation. If the numbers of iterations required are denoted by n_1 and n_2 ›spectively, it follows from (9.2) that

$$(p_1^{n_1} - p_2^{n_2})k_0 = -\left(\frac{p_1^{n_1}-1}{1-p_1}\right)\log_{10} C_1 + \left(\frac{p_2^{n_2}-1}{1-p_2}\right)\log_{10} C_2. \qquad (9.4)$$

One of the few examples in which it is possible to solve (9.4) for the ratio n_1/n_2 when the methods being compared are Newton's method and the secant method; ıen $p_N = 2, p_S = \frac{1}{2}(1+\sqrt{5}), C_N = \frac{1}{2}\{f''(\alpha)/f'(\alpha)\}$ and $C_S = C_N^{p-1}$. Thus ıbstituting in (9.4) it follows that

$$(2^{n_N} - p_S^{n_S})k_1 = (2^{n_N}-1)\log_{10} C_N - (p_S^{n_S}-1)\log_{10} C_N$$

ıd hence

$$2^{n_N} - p_S^{n_S} = 0;$$

ıat is,

$$n_S = n_N \frac{\log 2}{\log p_S} = n_N 1.44. \qquad (9.5)$$

Thus the secant method in general requires 1.44 times as many iterations as the ›ewton to achieve a given accuracy. The Newton method however requires the ılue of $f(x^{(n)})$ and $f'(x^{(n)})$ at each step, whereas the secant method only requires (x_n) to be computed. If it is assumed that the cost (in computer time) of ›aluating $f(x)$ is K and the cost of evaluating $f'(x)$ is $\gamma_1 K$, then the ratio of the ›sts to achieve the same accuracy with the two different methods, is given by

$$\frac{n_S K}{n_N(1+\gamma_1)K} = \frac{1.44}{1+\gamma_1}.$$

Thus,

secant is cheaper/quicker when $\gamma_1 > 0.44$;
Newton is cheaper/quicker when $\gamma_1 < 0.44$.

Jarratt (1970) has shown that it is possible to compare Newton and secant iteratic with certain multistage methods for which the asymptotic error constant is a function of $C_N = \frac{1}{2}\{f''(\alpha)/f'(\alpha)\}$.

*9.2 More General Comparisons

For comparing methods when no simple relation exists between the asymptoti error constants, various authors have suggested that the right-hand side of (9.4) is replaced by zero and thus

$$\frac{n_1}{n_2} = \frac{\log_{10} p_2}{\log_{10} p_1}. \qquad (9.$$

This provides a very straightforward criterion for methods when the function values $f(x^{(n)})(n = 0, 1, \ldots)$ alone are to be used. The order increases with the number of points used in the formula and so by this test – that is according to (9.6) – two-step methods such as Muller's method are preferable to one-step methods such as the secant method. The increase in order obtained by using k-ste methods with $k \geqslant 3$ is marginal as the order cannot exceed 2; in addition the formulae become increasingly complex. Similarly it is possible to show that the modification of Newton's method in which an old value of the derivative is retaine is not a particularly efficient method for a single equation, even when Δ^2-accelera is used at each step. From Section 8.3 it follows that it is possible to devise a fifth-order three-stage method from (8.14) if Aitken's Δ^2-acceleration is included (see Exercise 8.8). Each step of such a method involves three function evaluations and one derivative; thus the cost of n steps is $(3 + \gamma_1)Kn$, and hence using (9.6) it estimated that to achieve the same accuracy as Muller's method increases the cost by the factor

$$\frac{(3 + \gamma_1)\log_{10} 1.839}{\log_{10} 5} > 1.2, \qquad \gamma_1 \geqslant 0. \qquad (9.$$

Jarratt (1970) has shown that there exists a three-stage method of order five that involves one function and three derivative evaluations. In this case clearly the ratio (9.7) is replaced by

$$\frac{(1 + 3\gamma_1)\log_{10} 1.839}{\log_{10} 5},$$

which is less than one when $\gamma_1 < 0.547$. The reference contains a list of types of methods that could be chosen depending on the value of γ_1.

Algorithms have been devised which incorporate a strategy for selecting the most efficient method for a particular problem; see for example Nesdore (1970). Once the type of method – say two-step, order 1.839, using function value only – has been fixed, it is necessary to select a particular formula of that type. In Chapter 5, several two-step methods were mentioned: (5.4) and (5.5) were derived from Newton's method (5.9) by direct interpolation, (5.10) by inverse interpolation, and (5.11) by rational interpolation. Jarratt has suggested that if information is available on the behaviour of $f(x)$ in the neighbourhood of the root, this second selection can be based on the form of the asymptotic error constants.

Dowell and Jarratt (1972) have shown, subject to certain conditions on the behaviour of $f(x)$ in the neighbourhood of the solution, that asymptotically the Pegasus method produces a sequence

$$\ldots UU\,MM\;UU\,MM \ldots,$$

where U represents a step at which the secant iteration is used and M represents a step at which the modification procedure has to be used. This result clearly does not cover pathological cases such as Exercise 6.13. They have further shown that the index I_1 for the Pegasus method is 1.642 and hence it is possible to introduce this method into the comparison.

The comparison of methods for systems is considerably more complex, and such comparisons as exist (Murray, 1972) are largely empirical in nature.

Historical Appendix

An historical note on the iteration now known as Newton's method; comparing the contributions of Sir Isaac Newton and J. R. Raphson, Esq., based on '*Tracts on the resolution of affected algebraic equations*' by F. Maseres (1800).

Newton proposed the method in 1666 in *Analysis per Aequationes Numero Terminorum Infinitas.*

Newton's Example Find the roots of $x^3 - 2x = 5$.

Take $x = 2$ as a first approximation; the root lies between 2 and 3 since if $f(x) \equiv x^3 - 2x - 5$, then

$$f(2) = 8 - 4 - 5 = -1 \qquad \text{and} \qquad f(3) = 27 - 6 - 5 = 16.$$

Assume the root is $2 + z$; then

$$0 = f(2+z) \equiv (2+z)^3 - 2(2+z) - 5 = z^3 + 6z^2 + 10z - 1 \equiv f_1(z) \quad \text{(say)}.$$

Retaining only linear terms leads to $z = \frac{1}{10}$. Since

$$f_1(z) = 0 \Rightarrow z = \frac{1}{10} - \frac{6z^2 + z^3}{10},$$

it follows that root is less than $2\frac{1}{10}$. As the linear and constant terms in $f_1(z)$ correspond to f' and f evaluated at $x = 2$, using *synthetic division*, it follows that the correction is

$$-\frac{f(2)}{f'(2)}.$$

Raphson's Method (proposed in *Analysis Aequationum Universalis* (1690))
Assume solution is $2.1 - v$; then

$$\begin{aligned} 0 = f(2.1 - v) &\equiv (2.1 - v)^3 - 2(2.1 - v) - 5 \\ &= 0.061 - 11.23v + 6.3v^2 - v^3 \\ &= f_2(v). \end{aligned}$$

Retaining only linear terms in $f_2(v) = 0$ leads to the correction

$$v = \frac{0.061}{11.23} \left(\equiv \frac{f(2.1)}{f'(2.1)} \right) .$$
$$= 0.0054,$$

nd the iteration proceeds in a like manner, for the next approximation is 2.0946, which is too small since $f_2(v)$ implies

$$v = \frac{0.061}{11.23} - \frac{6.3v^2 - v^3}{11.23} .$$

Newton's original method was to compute the sequence in a different manner, which if followed today, would lead to a very unstable algorithm. The first correction is computed from $f_1(z) = 0$; retaining only linear terms, the next correction is found as an improved approximate solution of $f_1(z) = 0$, rather than of $f(x) = 0$. Thus assume $z = 0.1 - v$; then

$$\begin{aligned} 0 = f_1(0.1 - v) &= (0.1 - v)^3 + 6(0.1 - v)^2 + 10(0.1 - v) - 1 \\ &= 0.061 - 11.23v + 6.3v^2 + v^3 \\ &\equiv f_2(v). \end{aligned}$$

Retaining linear terms only clearly leads to the same correction as in the Raphson method. The next step of the Newton method is then to linearize and set

$$f_2(0.0054 + w) = 0.$$

The solution obtained is $x = 2.09455148$. The judgement of Maseres in 1800 was that Newton's method led to a slight simplification in the arithmetic, whereas Raphson's method was a more straightforward concept and was to be preferred. He also observes the necessity for a good first approximation. This same equation, $x^3 - 2x - 5 = 0$, was used by Fourier in 1818 in his paper

'Question D'analyse Algebrique',

in *Bulletin des Sciences par la Société Philomathique* (pp. 61–7). He was able to use calculus to show that the iterations

$$x^{(n+1)} = x^{(n)} - \frac{f(x^{(n)})}{f'(x^{(n)})}$$

and

$$y^{(n+1)} = y^{(n)} - \frac{f(y^{(n)})}{f'(x^{(n)})}$$

lead to bounds on the solution of $f(x) = 0$ and that

$$x^{(0)} = 2.09455$$

and

$$y^{(0)} = 2.09456$$

lead to

$$x^{(1)} = 2.0945514815$$
$$y^{(1)} = 2.0945514816.$$

Maseres points out that in 1646, Vieta suggested piecewise linear approximations i

De numerosa Potestatum Adfectarum Resolutione,

but he failed to provide a suitable method of improving the approximation.

Both Maseres and Halley mention that the third-order process (5.26) now know as Halley's method was first introduced by a Monsieur de Lagny to solve equations of the form

$$x^m = N.$$

No dates or references are given by either source.

ibliography

tken, A. C. (1926). *Proc. Roy. Soc.*, Edin., **A46**, 289–305.
ıderson, N. and Björck, A. (1973). *B.I.T.*, **13**, 253–64.
ırd, Y. (1968). *Math. Comp.*, **22**, 665–6.
ırnes, J. (1965). *Comp. J.*, **8**, 66–72.
örck, A. (1967a). *B.I.T.*, **7**, 1–21.
örck, A. (1967b). *B.I.T.*, **7**, 257–78.
ameller, A., Allan, R. N., and Haman, Y. M. (1976). *Sparsity*, Pitman, London.
own, K. M. (1967). *Comm. A.C.M.*, **10**, 728–9.
oyden, C. G. (1965). *Math. Comp.*, **19**, 577–93.
oyden, C. G. (1968). *Comp. J.*, **12**, 94–9.
oyden, C. G. (1971). *Math. Comp.*, **25**, 285–94.
ınch, J. R. and Rose, D. J. (1976) (eds.). *Sparse Matrix Computations*, Academic Press, New York.
ıs, J. C. P. and Dekker, T. J. (1975). *A.C.M. Trans. Math. Software*, **1**, 330–45.
ıuchy, A. (1847). *Comptes Rendus*, **25**, 536.
ıuchy, A. (1882). *Oeuvres Complètes* (2nd Series) **4**, 573–609, Gauthiers-Villars, Paris.
ɔx, M. G. (1970). *Comp. J.*, **13**, 101–2.
ırtis, A. R. (1972) in Frieman, C. V. (ed.) *Information Processing '71*, North-Holland, Amsterdam.
ırtis, A. R. (1974). *J. Inst. Math. Applics.*, **12**, 121–6.
ırtis, A. R. and Reid, J. K. (1972). *J. Inst. Math. Applics.*, **10**, 118–24.
ıthill, E. and McKee, J. (1969). *Proc A.C.M. 2nd National Conf.*, 1
avidenko, D. F. (1953). *Dokl. Acad. Nauk. SSSR.*, **88**, 601–2.
avies, M. and Dawson, B. (1975). *Math. Comp.*, **24**, 133–5.
ennis, J. R. (1971). *Math. Comp.*, **25**, 559–67.
owell, M. and Jarratt, P. (1971). *B.I.T.*, **11**, 168–74.
owell, M. and Jarratt, P. (1972). *B.I.T.*, **12**, 503–8.
uff, I. S. (1976). *Report CSS-28*, A.E.R.E., Harwell.
uff, I. S. and Reid, J. K. (1974). *J. Inst. Math. Applics.*, **14**, 281–91.
ngeli, M., Ginsburg, Th., Rutishauser, H., and Stiefel, E. (1959). *Refined Iterative Methods for Computation of the Solution and the Eigenvalues for Self-Adjoint Boundary Value Problems*, Birkhäuser-Verlag, Basel.
letcher, R. (1970) in Rabinowitz (1970).
letcher, R. and Powell, M. J. D. (1963). *Comp. J.*, **6**, 163–8.
orsythe, G. E. and Moler, C. B. (1967). *Computer Solution of Linear Algebraic Equations*, Prentice-Hall, Englewood Cliffs.

Fröberg, C-E. (1970). *An Introduction to Numerical Analysis*, 2nd Ed. Addison Wesley, Reading, Mass.

Gear, W. B. (1971). *Numerical Initial Valve Problems in Ordinary Differential Equations*, Prentice-Hall, Englewood Cliffs.

Gentleman, M. (1973). *J. Inst. Math. Applics.*, **12**, 329–36.

George, J. A. (1976). *Sparse Matrix Techniques*, Lecture Notes, 572. Springer Verlag, Berlin, 1.

Gibbs, N. E., Poole, W. G. and Stockmeyer, P. K. (1976). *S.I.A.M. J. Num. Anal.*, **13**, 236–50.

Gill, P. E. and Murray, W. (1972). *J. Inst. Math. Applics.*, **9**, 91–108.

Goldstein, A. A. (1967). *Constructive Real Analysis*, Harper Row, New York.

Golub, G. (1965). *Numer. Math.*, **7**, 206–16.

Gourlay, A. R. and Watson, G. A. (1973). *Computational Methods for Matrix Eigenproblems*, Wiley, London.

Halley, E. (1694). *Phil. Trans Roy. Soc. London*, **18**, 136–45.

Henrici, P. (1964). *Elements of Numerical Analysis*. McGraw-Hill, New York.

Householder, A. S. (1970). *The Numerical Treatment of a Single Nonlinear Equation*, McGraw-Hill, New York.

Jacobs, D. A. H. (1977). (ed.) *The State of the Art in Numerical Analysis*, Academ Press, London.

Jarratt, P. (1966). *Comp. J.*, **9**, 304–7.

Jarratt, P. (1970) in Rabinowtz (1970).

Jennings, A. (1966). *Comp. J.*, **9**, 281–5.

Kowalik, J. and Osborne, M. R. (1968). *Methods for Unconstrained Optimisation Problems*, Elsevier, New York.

Lambert, J. L. (1973). *Computational Methods in Ordinary Differential Equations*, Wiley, London.

Lawson, C. L. and Hanson, R. J. (1974). *Solving Least Squares Problems*, Prentice-Hall, Englewood Cliffs.

Levy, H. and Lessman, F. (1959). *Finite Difference Equations*, Pitman, London.

Lill, S. A. (1976), in L. C. W. Dixon (ed.), *Optimisation in Action*, Academic Press, London.

Liu, W. H. and Sherman, A. H. (1976). *S.I.A.M. J. Num. Anal.*, **13**, 198–213.

Lootsma, F. A. (1972) (ed.). *Numerical Methods for Non-linear Optimisation*, Academic Press, London.

Markowitz, H. M. (1957). *Manag. Sci.*, **3**, 255–69.

Marquardt, D. W. (1963). *J.S.I.A.M.*, **11**, 431–41.

Mitchell, A. R. (1969). *Computational Methods in Partial Differential Equations*, Wiley, London.

Mitchell, A. R. and Wait, R. (1977). *The Finite Element Method in Partial Differential Equations*, Wiley, London.

Muller, D. E. (1956). *M.T.A.C.*, **10**, 208–15.

Murray, W. (1972) (ed.). *Numerical Methods for Unconstrained Optimisation*, Academic Press, London.

Nesdore, P. F. (1970) in Rabinowitz (1970).

Ortega, J. M. (1972). *Numerical Analysis: A Second Course*, Academic Press, New York.

Ortega, J. M. and Rheinboldt, W. C. (1970). *Iterative Solution of Nonlinear Equations in Several Variables*, Academic Press, New York.

Ostrowski, A. M. (1966). *Solution of Equations and Systems of Equations*, Academic Press, New York.

Powell, M. J. D. (1970) in Rabinowitz (1970).

owell, M. J. D. and Reid, J. K. (1969) in Morrell, A. J. H. (ed.), *Information Processing '68*, North-Holland, Amsterdam.
abinowitz, P. (1970) (ed.). *Numerical methods for non-linear algebraic equations*, Gordon and Breach, London.
eid, J. K. (1971a) (ed.). *Large Sparse Sets of Linear Equations*, Academic Press, London.
eid, J. K. (1971b) in Reid (1971a).
heinboldt, W. C. (1974). *Methods for Solving Systems of Nonlinear Equations*, Regional Conference series monograph, SIAM, Philadelphia.
ice, J. R. (1964). The *Approximation of Functions*, Vol. 1, Addison Wesley, Reading, Mass.
ice, J. R. (1966). *Math. Comp.*, **20**, 325–8.
ose, D. J. and Willoughby, R. A. (1972) (eds.). *Sparse Matrices and Their Applications*, Plenum Press, New York.
chröder, E. (1870). *Math. Annal.*, **2**, 317–65.
chubert, L. K. (1970). *Math. Comp.*, **24**, 27–30.
teffensen, J. F. (1933). *Skand. Aktua. Tidskr.* **16**, 64–72.
ornheim, L. (1964). *J.A.C.M.*, **11**, 210–20.
ewarson, R. P. (1970). *S.I.A.M. Rev.*, **12**, 527–43.
osovic, L. B. (1973). *S.I.A.M. J. Appld. Math.*, **25**, 142–8.
raub, J. (1964). *Iterative Methods for the Solution of Nonlinear Equations*, Prentice-Hall, Englewood Cliffs.
arga, R. S. (1962). *Matrix Iterative Analysis*, Prentice-Hall, Englewood Cliffs.
endroff, B. (1966). *Theoretical Numerical Analysis*, Academic Press, New York.
hittaker, E. T. (1918). *Proc. Math. Soc. Edin.*, **36**, 103–6.
ilkinson, J. H. (1961). *J.A.C.M.*, **8**, 281–330.
ilkinson, J. H. (1963). *Rounding Error in Algebraic Processes*, H.M.S.O., London.
ilkinson, J. H. (1965a). *The Algebraic Eigenvalue Problem*, O.U.P., Oxford.
ilkinson, J. H. (1965b). *Comp. J.*, **8**, 77–84.
ilkinson, J. H. (1976) in Jacobs (1977).
ilkinson, J. H. and Reinsch, C. (1971). *Handbook for Automatic Computation II, Linear Algebra*, Springer-Verlag, Berlin.
ynn, P. (1956). *Proc. Camb. Phil. Soc.*, **52**, 663–71.
oung, D. M. (1971). *Iterative solution of large linear systems*, Academic Press, New York.

ndex